SCUOLA NORMALE SUPERIORE

QUADERNI

C. LAURENT-THIÉBAUT
J. LEITERER

Andreotti-Grauert Theory on Real Hypersurfaces

PISA - 1995

ISBN: 978-88-7642-244-7

NOTE

This volume collects two articles by Christine Laurent-Thiébaut and Jürgen Leiterer which were submitted to, and accepted for publication by the Annali della Scuola Normale Superiore, Classe di Scienze.

Owing to the character of the systematic exposition of the new scientific results achieved and to the size of the work (which would have inevitably interfered with the backlog of the Annali), the authors agreed to have the two papers published as a separate volume. To speed up publication it was also decided to have them appear in the form they ware submitted.

CONTENTS

CHAPTER I

The q-convex case

I.0 - Introduction

The results of this work on the Andreotti-Grauert theory on real hypersurfaces have been announced in [24].

Let M be a real C^2-hypersurface in some n-dimensional complex manifold X the Levi form of which has at least q positive and at least q negative eigenvalues, $0 \leq q \leq (n-1)/2$. In this chapter we study the tangential Cauchy-Riemann equation on M from the viewpoint of uniform estimates and by means of integral formulas of the Grauert/Lieb-Henkin type. This method was first applied by Henkin and Airapetjan/Henkin to the tangential Cauchy-Riemann equation (see [9], [10], [11], [2]). Concerning other methods in the theory of the tangential Cauchy-Riemann equation we refer to the survey article [12] of Henkin and the more recent works of Hill/Nacinovich [16], [17] and Trèves [30].

In this chapter we restrict ourselves to forms on M which are of maximal holomorphic degree. This is no true restriction, since we admit forms with values in holomorphic vector bundles. The advantage of this approach is that forms of maximal holomorphic degree on M are true differential forms (not equivalence classes) and the tangential Cauchy-Riemann operator then is the usual exterior differential operator d on M (see Chapter II.1 in [30], where this is explained in more detail).

Hence, we can write the tangential Cauchy-Riemann equation on M in the form

$$(0.1) \qquad\qquad du = f, \quad df = 0,$$

where f and u are differential forms of maximal holomorphic degree with values in some holomorphic vector bundle over X.

Pervenuto alla Redazione il 14 Luglio 1994.

It is well-known ([3], [27], [10], [12], [30]) that this equation has local solutions in two cases: 1. f is of bidegree (n, r) with $n - q \leq r \leq n - 1$; 2. f is of bidegree (n, r) with $1 \leq r \leq q - 1$. In the present chapter the case $n - q \leq r \leq n - 1$ is studied. This case is called also the "convex case" reflecting the fact that for forms f of high degree equation (0.1) can be solved near to boundaries satisfying certain convexity conditions.

To describe the content of the chapter we want to emphasize two main results, a local and a global one.

The main local result is a local homotopy formula

$$(0.2) \qquad\qquad f = dTf - Tdf$$

for forms f on M of bidegree (n, r) with $r \geq n - q$, where the operator T admits Hölder estimates up to the boundary of a strictly q-convex domain $D \subset\subset M$ (Definition 5.1). Constructing this formula we follow the classical concept first used by Andreotti/Hill (see [4], [5]) for the study of the tangential Cauchy-Riemann equation on hypersurfaces. This concept consists in two steps: 1. Reduction of the problem by means of a jump formula to a $\bar{\partial}$-problem with estimates for certain auxiliary domains in $\mathbb{C}^n$; 2. Solving this $\bar{\partial}$-problem. We do this by means of integral operators of the Grauert/Lieb-Henkin type. The most important among these estimates are the Hölder estimates for piecewise strictly q-convex domains in $\mathbb{C}^n$ finally proved in [21] and going back to ideas of Henkin and Airapetjan/Henkin (cf. the introduction to [21]). Also we use a special uniform estimate for $\bar{\partial}$ proved in [8].

Without estimates up to a boundary inside M, but under weaker convexity condition on M and in the C^∞-case, local homotopy formulas for the tangential Cauchy-Riemann equation are constructed by Trèves [30] using integral operators of the Fourier-Bros-Iagolnitzer type.

Our homotopy operator in particular gives a local solution of (0.1) with Hölder estimates up to a strictly q-convex boundary in M if f of bidegree (n, r) with $r \geq n - q$. A corresponding C^∞-result (up to such a boundary) was proved in 1988 by Nacinovich [28]. In the special case when this boundary is the intersection of M with some strictly pseudoconvex boundary in X, the existence of such a solution was announced in 1981 by Henkin [10] (see also Theorem 8.15 in the survey article [12]).

The essence of the main global result (Section 8) can be stated as follows: let $D \subset\subset \Omega \subset\subset M$ be two q-convex open sets (Definition 7.1 (iii)) such that $D = \{\varphi < 0\}$ and $\Omega = \{\varphi < 1\}$, where φ is a real C^2-function on M which is $(q + 1)$-convex (Definition 2.4) on $\Omega \backslash D$, and let E be a holomorphic vector bundle on X. Then the restriction maps

$$(0.3) \qquad\qquad H^{n,r}(\Omega, E) \to H^{n,r}(D, E), \quad n - q \leq r \leq n - 1,$$

are isomorphisms.

The main global result will be deduced from the homotopy formula (0.2) by means of a version of Grauert's bump method. This is very close to the procedure used in Section 12 of the book [15] to prove the corresponding statement on complex manifolds. The main difference is that now the geometrical picture is more complicated – instead of the simple Lemma 12.3 in [15] here we have Theorem 7.10.

The fact that (0.3) is an isomorphism together with the local result mentioned above yields finiteness and vanishing theorems (Section 8). In the C^∞-case and without estimates up to a boundary inside M, but for CR-manifolds of arbitrary codimension, the corresponding finiteness and vanishing theorems were recently obtained by Hill and Nacinovich [16], [17] using different methods.

I.1 - Notations

Let X be an n-dimensional complex manifold and E a holomorphic vector bundle on X.

I.1.1 – Let D be an open subset with C^2-boundary in X, and let $\overline{D}$ be the closure of D in X.

We denote by $C^\alpha_{n,r}(\overline{D}, E)$ ($0 \le r \le n$, $0 \le \alpha < 1$) the space of continuous (if $\alpha = 0$), resp. Hölder continuous with exponent α (if $\alpha > 0$), E-valued differential forms of bidegree (n, r) on $\overline{D}$.

If $\overline{D}$ is compact, then $C^\alpha_{n,r}(\overline{D}, E)$ will be considered as *Banach space* endowed with the max-norm (if $\alpha = 0$), resp. the Hölder norm with exponent α (if $\alpha > 0$).

If $\overline{D}$ is not compact, then $C^\alpha_{n,r}(\overline{D}, E)$ will be considered as *Fréchet space* endowed with the topology defined by the Banach spaces $C^\alpha_{n,r}(\overline{W}, E)$ where W runs over all open sets $W \subseteq D$ with C^2-boundary such that the closure $\overline{W}$ of W in $\overline{D}$ is compact.

The forms which are Hölder continuous with exponent $1/2 - \varepsilon$ for all $\varepsilon > 0$ are of particular interest in this chapter. Therefore we introduce also the spaces

$$C^{<1/2}_{n,r}(\overline{D}, E) := \bigcap_{\varepsilon>0} C^{1/2-\varepsilon}_{n,r}(\overline{D}, E).$$

These spaces will be considered as Fréchet spaces endowed with the topology defined by the topologies of the spaces $C^{1/2-\varepsilon}_{n,r}(\overline{D}, E)$, $\varepsilon > 0$, *i.e.* a map with values in $C^{<1/2}_{n,r}(\overline{D}, E)$ is continuous if and only if it is continuous as a map with values in each $C^{1/2-\varepsilon}_{n,r}(\overline{D}, E)$, $\varepsilon > 0$.

We denote by $Z^\alpha_{n,r}(\overline{D}, E)$ ($0 \le r \le n$, $0 \le \alpha < 1$) and $Z^{<1/2}_{n,r}(\overline{D}, E)$ ($0 \le \alpha < 1$) the subspaces of closed forms in $C^\alpha_{n,r}(\overline{D}, E)$ resp. $C^{<1/2}_{n,r}(\overline{D}, E)$. For $1 \le r \le n$ we set

$$E^{<1/2}_{n,r}(\overline{D}, E) = Z^0_{n,r}(\overline{D}, E) \cap dC^{<1/2}_{n,r-1}(\overline{D}, E)$$

and

$$H^{n,r}_{<1/2}(\overline{D}, E) = Z^0_{n,r}(\overline{D}, E) / E^{<1/2}_{n,r}(\overline{D}, E).$$

It will be convenient to use also the corresponding spaces of *germs*: for $r = 0, \ldots, n$ and $0 \le \alpha < 1$, we denote by germ $C_{n,r}^{\alpha}(\overline{D}, E)$ the space of germs of continuous (n, r)-forms in neighborhoods of $\overline{D}$ which are Hölder continuous with exponent α if $\alpha > 0$, and we set

$$\text{germ } C_{n,r}^{<1/2}(\overline{D}, E) = \bigcap_{\varepsilon > 0} \text{germ } C_{n,r}^{1/2 - \varepsilon}(\overline{D}, E),$$

$$\text{germ } Z_{n,r}^{\alpha}(\overline{D}, E) = \{f \in C_{n,r}^{\alpha}(\overline{D}, E): df = 0 \ (\text{as a germ})\},$$

$$\text{germ } Z_{n,r}^{<1/2}(\overline{D}, E) = \bigcap_{\varepsilon > 0} \text{germ } Z_{n,r}^{1/2 - \varepsilon}(\overline{D}, E).$$

For $r = 1, \ldots, n$ we set

and

$$\text{germ } E_{n,r}^{<1/2}(\overline{D}, E) = \text{germ } Z_{n,r}^{0}(\overline{D}, E) \cap d \text{ germ } C_{n,r-1}^{<1/2}(\overline{D}, E)$$

$$\text{germ } H_{<1/2}^{n,r}(\overline{D}, E) = \text{germ } Z_{n,r}^{0}(\overline{D}, E) / \text{germ } E_{n,r}^{<1/2}(\overline{D}, E).$$

We do *not* introduce topologies in spaces of germs, because we do not need this.

I.1.2 – Let M be a real C^2-hypersurface in X (not necessarily closed) and D an open subset with C^2-boundary in M, and let $\overline{D}$ be the closure of D in M.

A continuous E-valued differential form on D will be called of *bidegree* (n, r), $0 \le r \le n - 1$, if it is the restriction to D of a continuous E-valued (n, r)-form defined in some X-neighborhood of D. Using this definition, in the same way as in Section 1.1 we define the spaces: $C_{n,r}^{\alpha}(\overline{D}, E)$, $Z_{n,r}^{\alpha}(\overline{D}, E)$ $(0 \le r \le n - 1, \ 0 \le \alpha < 1)$; $C_{n,r}^{<1/2}(\overline{D}, E)$, $Z_{n,r}^{<1/2}(\overline{D}, E)$ $(0 \le r \le n - 1)$; $E_{n,r}^{<1/2}(\overline{D}, E)$, $H_{<1/2}^{n,r}(\overline{D}, E)$ $(1 \le r \le n - 1)$; germ $C_{n,r}^{\alpha}(\overline{D}, E)$, germ $Z_{n,r}^{\alpha}(\overline{D}, E)$ $(0 \le r \le n - 1, \ 0 \le \alpha < 1)$, germ $C_{n,r}^{<1/2}(\overline{D}, E)$, germ $Z_{n,r}^{<1/2}(\overline{D}, E)$ $(0 \le r \le n - 1)$; germ $E_{n,r}^{<1/2}(\overline{D}, E)$, germ $H_{<1/2}^{n,r}(\overline{D}, E)$ $(1 \le r \le n - 1)$.

I.1.3 – If we write in the definitions above a $*$ instead of r then we mean the union over all r of the corresponding spaces. For example, $C_{n,*}^{\alpha}(\overline{D}, E)$ is the union of all $C_{n,r}^{\alpha}(\overline{D}, E)$ with $0 \le r \le n$ (if D is an open subset of X) resp. $0 \le r \le n - 1$ (if $D \subseteq M$).

I.1.4 – If E is the trivial line bundle, then we omit the bundle E in the definitions above, *i.e.* we write $C_{n,r}^{\alpha}(\overline{D})$ instead of $C_{n,r}^{\alpha}(\overline{D}, E)$ etc.

I.1.5 – For $\xi \in X$ we denote by $T_{\xi}^{1,0}(X)$ the holomorphic tangent space of X at ξ, *i.e.* if $z_1, \ldots, z_n$ are holomorphic coordinates in a neighborhood of ξ, then

$T_\xi^{1,0}(X)$ consists of all tangent vectors t of the form

$$(1.1) \qquad t = \sum_{j=1}^{n} t_j \frac{\partial}{\partial z_j}(\xi)$$

where $t_1, \ldots, t_n$ are complex numbers.

If M is a real C^2-submanifold of X and $\xi \in M$, then we denote by $T_\xi^{1,0}(M)$ the subspace of all vectors in $T_\xi^{1,0}(X)$ which are tangential for M.

I.1.6 – Let $\xi \in X$ and ρ a real C^2-function defined in a neighborhood of ξ. Then we denote by $L_\xi^X(\rho)$ the Levi form of ρ at ξ, *i.e.* the hermitian form on $T_\xi^{1,0}(X)$ defined by

$$L_\xi^X(\rho)t = \sum_{j,k=1}^{n} \frac{\partial^2 \rho(\xi)}{\partial z_j \partial \bar{z}_k} t_j \bar{t}_k$$

if $t \in T_\xi^{1,0}(X)$ is written in the form (1.1). If M is a real C^2-submanifold of X and $\xi \in M$, then we denote by $L_\xi^M(\rho)$ the restriction of $L_\xi^X(\rho)$ to $T_\xi^{1,0}(M)$.

I.2 - $(q+1)$-convex functions

First recall the following well-known proposition (cf., e.g. Sections 4 and 5 in [15]).

PROPOSITION 2.1. *Let M be a real C^2-hypersurface in some n-dimensional complex manifold X, let $\xi \in M$, and let q be an integer with $0 \le q \le (n-1)/2$. Then the following conditions are equivalent:*

(i) *There exist real C^2-functions ρ_+ and ρ_- in some X-neighborhood U of ξ such that $M \cap U = \{\rho_+ = 0\} = \{\rho_- = 0\}$, $d\rho_\pm(\xi) \neq 0$, $\rho_+(\varsigma)\rho_-(\varsigma) < 0$ if $\varsigma \in U \backslash M$, and the Levi forms $L_\xi^M(\rho_\pm)$ have at least q positive eigenvalues.*

(ii) *There exist real C^2-functions ρ_+ and ρ_- in some X-neighborhood U of ξ such that $M \cap U = \{\rho_+ = 0\} = \{\rho_- = 0\}$, $d\rho_\pm(\xi) \neq 0$, $\rho_+(\varsigma)\rho_-(\varsigma) < 0$ if $\varsigma \in U \backslash M$, and the Levi forms $L_\xi^X(\rho_\pm)$ have at least $q+1$ positive eigenvalues.*

(iii) *For each real C^2-function ρ in some X-neighborhood U of ξ with $M \cap U = \{\rho = 0\}$ and $d\rho(\xi) \neq 0$, $L_\xi^M(\rho)$ has at least q positive and q negative eigenvalues.*

DEFINITION 2.2. Let M be a real C^2-hypersurface in some n-dimensional complex manifold X and let q be an integer with $0 \le q \le (n-1)/2$. M will be called *q-convex-concave at a point* $\xi \in M$ if the equivalent conditions (i)-(iii) in Proposition 2.1 are fulfilled. M will be called *q-convex-concave* if it is q-convex-concave at each point in M.

PROPOSITION 2.3. *Let M be a real C^2-hypersurface in some n-dimensional complex manifold X, q an integer with $0 \leq q \leq (n-1)/2$, and φ a real C^2-function on M. Then, for each point $\xi \in M$, the following two conditions are equivalent:*

(i) *There exist an X-neighborhood U of ξ and real C^2-functions $\tilde{\varphi}$, ρ_+, ρ_- on U with the following properties:*

- *$\tilde{\varphi} = \varphi$ on $U \cap M$;*
- *$M \cap U = \{\rho_+ = 0\} = \{\rho_- = 0\}$;*
- *$d\rho_\pm(\varsigma) \neq 0$ for all $\varsigma \in U$;*
- *$\rho_+\rho_- < 0$ on $U \backslash M$;*
- *for each $\lambda \in [0,1]$ and for each $\varsigma \in U$, the forms $L_\varsigma^X(\lambda\tilde{\varphi} + (1-\lambda)\rho_\pm)$ have at least $q+1$ positive eigenvalues.*

(ii) *M is q-convex-concave at ξ and, for any real C^2-extension ψ of φ to an X-neighborhood of ξ, $L_\xi^M(\psi)$ has at least q positive eigenvalues.*

PROOF. First assume that condition (i) is fulfilled. Setting $\lambda = 0$ then we see that M is q-convex-concave at ξ. Now let ψ be a real C^2-extension of φ to some X-neighborhood U of ξ. If U is sufficiently small, then there are real C^2-functions $\tilde{\varphi}$, ρ_+, ρ_- on U with the properties listed in condition (i). let φ_+ and φ_- be the functions on U with

$$\psi = \tilde{\varphi} + \varphi_\pm\rho_\pm.$$

Since $d\rho_\pm(t) = 0$ if $t \in T_\xi^{1,0}(M)$, then

(2.1) $$L_\xi^M(\psi) = L_\xi^M(\tilde{\varphi}) + \varphi_\pm(\xi)L_\xi^M(\rho_\pm).$$

Since $\rho_+\rho_- < 0$ on $U \backslash M$, at least one of the numbers $\varphi_+(\xi)$ and $\varphi_-(\xi)$ is non-negative. We may assume that this is $\varphi_+(\xi)$. Set

$$\lambda = \frac{1}{1 + \varphi_+(\xi)}.$$

Then $0 < \lambda \leq 1$ and it follows from (2.1) that

$$L_\xi^M(\psi) = \frac{1}{\lambda} L_\xi^M(\lambda\tilde{\varphi} + (1-\lambda)\rho_+).$$

Since $L_\xi^X(\lambda\tilde{\varphi} + (1-\lambda)\rho_+)$ has at least $q+1$ positive eigenvalues, this completes the proof of condition (ii).

Now we assume that condition (ii) is fulfilled. Since then M is q-convex-concave at ξ, then we can find real C^2-functions ρ_+ and ρ_- in some X-neighborhood W of ξ such that $M \cap W = \{\rho_+ = 0\} = \{\rho_- = 0\}$, $d\rho_\pm(\xi) \neq 0$, $\rho_+\rho_- < 0$ on $U \backslash M$, and the forms $L_{\rho_\pm}^X(\xi)$ have at least $q+1$ positive eigenvalues.

We choose W so small that φ is defined on $W \cap M$. Further we fix an arbitrary real C^2-function ψ on W with $\psi = \varphi$ on $W \cap M$. Set

$$\varphi_c = \psi + c\rho_+^2 \quad \text{for } c > 0.$$

To prove condition (i), now we prove the following more precise statement: there exist a constant $c_0 > 0$ and an X-neighborhood U of ξ such that, for all $c \geq c_0$, the functions $\tilde{\varphi} := \varphi_c$, ρ_+, ρ_- have the properties listed in condition (i). For the proof of this statement it remains to show that, for all $\lambda \in [0,1]$, the following assertion holds:

(2.2) $\left.\begin{array}{l}\text{There exists a constant } c_\lambda > 0 \text{ such that,}\\[4pt]\text{for all } c \geq c_\lambda, \text{ the forms } L_\xi^X(\lambda\varphi_c + (1-\lambda)\rho_\pm)\\[4pt]\text{have at least } q+1 \text{ positive eigenvalues.}\end{array}\right\}$

First we prove that for all $\lambda \in [0,1]$ the following holds:

(2.3) $\left.\begin{array}{l}\text{The forms } L_\xi^M(\lambda\psi + (1-\lambda)\rho_\pm) \text{ have at}\\[4pt]\text{least } q \text{ positive eigenvalues.}\end{array}\right\}$

If $\lambda = 0$ then (2.3) follows from the fact that the forms $L_\xi^X(\rho_\pm)$ have at least $q+1$ positive eigenvalues. If $\lambda = 1$ then (2.3) holds in view of condition (ii). Now let $0 < \lambda < 1$. Then we consider the real C^2-extensions $\varphi_\pm$ of φ to W defined by

$$\varphi_\pm = \psi + \frac{1-\lambda}{\lambda}\rho_\pm.$$

By condition (ii), the forms $L_\xi^M(\varphi_\pm)$ have at least q positive eigenvalues. Since $L_\xi(\lambda\psi + (1-\lambda)\rho_\pm) = \lambda L_\xi(\varphi_\pm)$, this completes the proof of (2.3).

Now we come to the proof of (2.2). Since the forms $L_\xi^X(\rho_\pm)$ have at least $q+1$ positive eigenvalues, it is clear that (2.2) holds for $\lambda = 0$. Let $0 < \lambda \leq 1$. By (2.3) there is a q-dimensional subspace T_q of $T_\xi^{1,0}(M)$ such that the forms $L_\xi^M(\lambda\psi + (1-\lambda)\rho_\pm)$ are positive definite on T_q. Choose a $(q+1)$-dimensional subspace T_{q+1} of $T_\xi^{1,0}(X)$ with $T_q = T_{q+1} \cap T_\xi^{1,0}(M)$. Let S_{q+1} be the unit sphere in T_{q+1} and denote by S_{q+1}^- the set of all $t \in S_{q+1}$ with

$$\min(L_\xi^X(\lambda\psi + (1-\lambda)\rho_+)t, \ L_\xi^X(\lambda\psi + (1-\lambda)\rho_-)t) \leq 0.$$

Since $T_q = \{t \in T_{q+1} : d\rho_+(t) = 0\}$ and the forms $L_\xi^X(\lambda\psi + (1-\lambda)\rho_\pm)$ are positive definite on T_q, then $d\rho_+(t) \neq 0$ if $t \in S_{q+1}^-$. Set

$$c_\lambda = \max_{t \in S_{q+1}^-} \frac{|L_\xi^X(\lambda\psi + (1-\lambda)\rho_+)t| + |L_\xi^X(\lambda\psi + (1-\lambda)\rho_-)t|}{|d\rho_+(t)|^2}.$$

Since $L_\xi^X(\rho_+^2)t = 2|d\rho_+(t)|^2$, then for all $c \geq c_\lambda$ and $t \in S_{q+1}^-$

$$L_\xi^X(\lambda\varphi_c + (1 - \lambda)\rho_\pm)t = L_\xi^X(\lambda\psi + (1 - \lambda)\rho_\pm)t + 2c|d\rho_+(t)|^2 > 0.$$

Together with the definition of S_{q+1}^- this implies that the forms $L_\xi^X(\lambda\varphi_c+(1-\lambda)\rho_\pm)$ are positive definite on T_{q+1} if $c \geq c_\lambda$. ∎

DEFINITION 2.4. Let M be a real C^2-hypersurface in an n-dimensional complex manifold X, and let q be an integer with $0 \leq q \leq (n-1)/2$. A real C^2-function φ on M will be called $(q+1)$-*convex at a point* $\xi \in M$ if the equivalent conditions (i) and (ii) in Proposition 2.3 are fulfilled. A $(q + 1)$-*convex function on M* is, by definition, a real C^2-function on M which is $(q+1)$-convex at all points in M.

REMARK 2.5. If M is a q-convex-concave real C^2-hypersurface in an n-dimensional complex manifold X, $0 \leq q \leq (n-1)/2$, and ψ is a strictly plurisubharmonic function on X, then the restriction of ψ to M is $(q+1)$-convex on M in the sense of the previous definition.

DEFINITION 2.6. Let M be a real C^2-hypersurface in some complex manifold X and let φ be a real C^2-function on M. Recall that a point $\xi \in M$ is called *critical for* φ if $d\varphi(\xi) = 0$. A point $\xi \in M$ will be called *generic for* φ if ξ is non-critical for φ (and hence $\varphi^{-1}(\varphi(\xi))$ is smooth in a neighborhood of ξ) and, moreover,

$$T_\xi^{1,0}(\varphi^{-1}(\varphi(\xi))) \neq T_\xi^{1,0}(M).$$

PROPOSITION 2.7. *Let M be a q-convex-concave real C^2-hypersurface in some n-dimensional complex manifold X, $1 \leq q \leq (n-1)/2$, and let φ be a real C^2-function on M satisfying for all $\xi \in M$ the following condition:*

(2.5) $\left.\begin{array}{l} \textit{if } \xi \textit{ is non-generic for } \varphi, \textit{ then } \varphi \textit{ is } (q+1)\textit{-convex at} \\[4pt] \xi, \textit{ and if } \xi \textit{ is generic for } \varphi, \textit{ then, for any real } C^2\text{-} \\[4pt] \textit{extension } \psi \textit{ of } \varphi \textit{ to some } X\text{-}\textit{neighborhood of } \xi, \\[4pt] L_\xi^{\varphi^{-1}(\varphi(\xi))}(\psi) \textit{ has at least } q - 1 \textit{ positive eigenvalues.} \end{array}\right\}$

Then, for each compact set $K \subset\subset M$, there exists $C_K > 0$ such that the following is true: if $\chi : \mathbb{R} \to \mathbb{R}$ is a C^2-function with

(2.6) $$\chi'(x) > 0 \text{ and } \chi''(x) \geq C_K\chi'(x) \text{ for all } x \in \varphi(K),$$

then $\chi \circ \varphi$ is $(q+1)$-convex at all points in K. (Example: $\chi(x) = \exp(Cx)$ where $C \geq C_K$.)

PROOF. Take real C^2-functions ρ and $\tilde\varphi$ in some X-neighborhood of M such that $M = \{\rho = 0\}$, $d\rho(\varsigma) \neq 0$ for $\varsigma \in M$, and $\tilde\varphi(\varsigma) = \varphi(\varsigma)$ for $\varsigma \in M$. Denote by

$H(\varsigma)$, $\varsigma \in M$, the positive semi-definite hermitian form on $T_\varsigma^{1,0}(M)$ defined by

$$H(\varsigma)t := |d\varphi(t)|^2, \quad t \in T_\varsigma^{1,0}(M).$$

Now we first prove the following auxiliary statement:

(2.7)
$$\left.\begin{array}{l} \text{There exists } C_K > 0 \text{ such that, for all } \varsigma \in K \\[4pt] \text{and } \lambda \in \mathbb{R}, \text{ the form } C_K H(\varsigma) + L_\varsigma^M(\tilde\varphi + \lambda\rho) \\[4pt] \text{has at least } q \text{ positive eigenvalues.} \end{array}\right\}$$

Since M is q-convex-concave, $L_\varsigma^M(\rho)$ has at least q positive and at least q negative eigenvalues. Therefore we can find $\lambda_K > 0$ such that, for all $\varsigma \in K$ and $\lambda \in \mathbb{R}$ with $|\lambda| \geq \lambda_K$,

$$L_\varsigma^M(\tilde\varphi + \lambda\rho) = L_\varsigma^M(\tilde\varphi) + \lambda L_\varsigma^M(\rho)$$

has at least q positive eigenvalues. Moreover, if $\varsigma \in M$ is non-generic for φ and $\lambda \in \mathbb{R}$, then $\tilde\varphi + \lambda\rho$ is a real C^2-extension of φ to some X-neighborhood of ς and, by hypothesis of the proposition, $L_\varsigma^M(\tilde\varphi + \lambda\rho)$ has at least q positive eigenvalues. Hence, for the proof of (2.7) it is sufficient to show that the following is true:

(2.8)
$$\left.\begin{array}{l} \text{Let } \xi \in M \text{ be generic for } \varphi, \text{ and let } \lambda_0 \in \mathbb{R}. \text{ Then} \\[4pt] \text{there exist } C, \ \varepsilon > 0 \text{ and a neighborhood } U \subseteq M \\[4pt] \text{of } \xi \text{ such that, for all } \varsigma \in U \text{ and } \lambda \in \mathbb{R} \text{ with} \\[4pt] |\lambda - \lambda_0| \leq \varepsilon, \text{ the form } C H(\varsigma) + L_\varsigma^M(\tilde\varphi + \lambda\rho) \\[4pt] \text{has at least } q \text{ positive eigenvalues.} \end{array}\right\}$$

We prove (2.8). Since $\tilde\varphi + \lambda_0\rho$ is a real C^2-extension of φ to some X-neighborhood of ξ, by hypothesis of the proposition, we can find a $(q-1)$-dimensional subspace T_{q-1} of $T_\xi^{1,0}(\varphi^{-1}(\varphi(\xi)))$ such that $L_\xi^X(\tilde\varphi + \lambda_0\rho)$ is positive definite on T_{q-1}. Since $T_\xi^{1,0}(\varphi^{-1}(\varphi(\xi))) \neq T_\xi^{1,0}(M)$, we can find a q-dimensional subspace T_q of $T_\xi^{1,0}(M)$ with

$$T_{q-1} = T_q \cap T_\xi^{1,0}(\varphi^{-1}(\varphi(\xi))).$$

Let S_q be the unit sphere in T_q and let

$$S_q^- := \{t \in S_q : L_\xi^M(\tilde\varphi + \lambda_0\rho)t \leq 0\}.$$

Then $H(\xi)t > 0$ for all $t \in S_q^-$. Set

$$C = 2 \max_{t \in S_q^-} \frac{|L_\xi^M(\tilde\varphi + \lambda_0\rho)t|}{H(\xi)t}.$$

Then $CH(\xi) + L_\xi^M(\tilde{\varphi} + \lambda_0\rho)$ is positive definite on S_q and has therefore at least q positive eigenvalues. By continuity, we can find $\varepsilon > 0$ and a neighborhood $U \subseteq M$ of ξ such that the same is true for $CH(\varsigma) + L_\varsigma^M(\tilde{\varphi} + \lambda\rho)$ if $|\lambda - \lambda_0| \leq \varepsilon$ and $\varsigma \in U$. This completes the proof of (2.8) and hence of (2.7).

Now we take the constant C_K from (2.7) and consider a C^2-function $\chi \colon \mathbb{R} \to \mathbb{R}$ satisfying condition (2.6). Let $\xi \in K$ and let μ be a real C^2-extension of $\chi \circ \varphi$ to an X-neighborhood of ξ. To complete the proof of the proposition, we have to show that $L_\xi^M(\mu)$ has a least q positive eigenvalues.

Since $\chi'(\varphi(\xi)) \neq 0$, we may assume that μ is of the form $\mu = \chi \circ \psi$, where ψ is a real C^2-extension of φ to an X-neighborhood of ξ. Then $d\psi(t) = d\varphi(t)$ for $t \in T_\xi^{1,0}(M)$. Hence

$$(2.9) \qquad L_\xi^M(\mu) = \chi''(\varphi(\xi))H(\xi) + \chi'(\varphi(\xi))L_\xi^M(\psi).$$

If ψ_1 is the function in an X-neighborhood of ξ with $\psi = \tilde{\varphi} + \psi_1\rho$ and if $\lambda := \psi_1(\xi)$, then $L_\xi^M(\psi) = L_\xi^M(\tilde{\varphi} + \lambda\rho)$. Together with (2.9) and (2.7) this implies that $L_\xi^M(\mu)$ has at least q positive eigenvalues. ∎

I.3 - Local jumps

In this section we work in $\mathbb{C}^n$ and denote by $B = B(z, \varsigma)$ the Bochner-Martinelli-Koppelman kernel

$$B(z, \varsigma) = \frac{(n-1)!}{(2\pi i)^n} \sum_{j=1}^n (-1)^{j+1} \frac{\bar{\varsigma}j - \bar{z}j}{|\varsigma - z|^{2n}} \bigwedge_{k \neq j} (d\bar{\varsigma}_k - d\bar{z}_k) \wedge \bigwedge_{k=1}^n (d\varsigma_k - dz_k).$$

If Ω is a compact oriented C^2-hypersurface with C^2-boundary $\partial\Omega$ in $\mathbb{C}^n$, then, for each continuous differential form f on Ω, resp. $\partial\Omega$, we use the abbreviations

$$(3.1) \qquad B_\Omega f(z) := \int_{\varsigma \in \Omega} f(\varsigma) \wedge B(z, \varsigma), \quad z \in \mathbb{C}^n \backslash \Omega,$$

and

$$(3.2) \qquad B_{\partial\Omega} f(z) := \int_{\varsigma \in \partial\Omega} f(\varsigma) \wedge B(z, \varsigma), \quad z \in \mathbb{C}^n \backslash \partial\Omega.$$

In the following we shall use the notation $\|f(\varsigma)\|$ to denote the *norm of the differential form* f *at the point* ς (cf., e.g., Section 0.4 in [15]). If M is a manifold in some open subset D_f of which f is defined, N is a submanifold of D_f and $f|_N$ is the restriction of f to N, then, for $\varsigma \in N$, we shall write also $\|f(\varsigma)\|_M$ instead of $\|f(\varsigma)\|$ and $\|f(\varsigma)\|_N$ instead of $\|f|_N(\varsigma)\|$.

For convenience of the reader now we collect in the following proposition some of the well-known properties of the Bochner-Martinelli-Koppelman integral.

PROPOSITION 3.1. *Let Ω be a compact oriented C^2-hypersurface with C^2-boundary $\partial\Omega$ in $\mathbb{C}^n$. Then:*

(i) *There exists a constant $C > 0$ such that*

(3.3) $$\|B_\Omega f(z)\|_{\mathbb{C}^n} \leq C(1 + |\ell n \operatorname{dist}(z,\Omega)|)\max_{\varsigma\in\Omega}\|f(\varsigma)\|_\Omega$$

for all $f \in C^0_{n,*}(\Omega)$ and $z \in \mathbb{C}^n\backslash\Omega$. Moreover, if $f \in C^\alpha_{n,*}(\Omega)$ for some $\alpha > 0$, then $B_\Omega f$ admits continuous extensions from both sides to $\Omega\backslash\partial\Omega$.

(ii) *Let ρ, φ be real C^2-functions on $\mathbb{C}^n$ with $\Omega = \{\rho = 0\} \cap \{\varphi \leq 0\}$, $d\rho(z) \neq 0$ for $z \in \Omega$, and $d\rho(z) \wedge d\varphi(z) \neq 0$ for $z \in \partial\Omega$. Set*

(3.4) $$\gamma(z) = \|\partial\rho(z) \wedge \partial\varphi(z)\|_{\mathbb{C}^n}.$$

Then there exists $C > 0$ such that

(3.5) $$\|B_{\partial\Omega}f(z)\|_{\mathbb{C}^n} \leq C\left(\frac{\gamma(z)}{\operatorname{dist}(z,\partial\Omega)} + 1 + |\ell n \operatorname{dist}(z,\partial\Omega)|\right)\max_{\varsigma\in\partial\Omega}\|f(\varsigma)\|_\Omega$$

for all $f \in C^0_{n,*}(\Omega)$ and $z \in \mathbb{C}^n\backslash\partial\Omega$.

(iii) *If $f \in C^0_{n,*}(\Omega)$ and $df \in C^0_{n,*}(\Omega)$, then*

(3.6) $$dB_\Omega f + B_\Omega df = B_{\partial\Omega}f \quad on \quad \mathbb{C}^n\backslash\Omega$$

and

(3.7) $$dB_{\partial\Omega}f - B_{\partial\Omega}df = 0 \quad on \quad \mathbb{C}^n\backslash\partial\Omega.$$

(iv) *Let $f \in C^0_{n,r}(\Omega)$ and $df \in C^0_{n,r+1}(\Omega)$, $0 \leq r \leq n - 1$. Then*

(3.8) $$(-1)^{n+1}\int_\Omega f \wedge \varphi = \int_{\mathbb{C}^n\backslash\Omega} dB_\Omega f \wedge \varphi + (-1)^{n+r}\int_{\mathbb{C}^n\backslash\Omega} B_\Omega f \wedge d\varphi$$

for all C^∞-forms φ with compact support in $\mathbb{C}^n\backslash\partial\Omega$. (The integrals on the right hand side of (3.8) exist in view of estimates (3.3), (3.5) and relation (3.6).) If, moreover, f is Hölder continuous on Ω and $B^+_\Omega f$, resp. $B^-_\Omega f$, is the continuous extension of $B_\Omega f$ from the left, resp. the right, to $\Omega\backslash\partial\Omega$ (which then exists by part (i) of the proposition), then

(3.9) $$(-1)^{n+r}f(z) = B^+_\Omega f(z)|_\Omega - B^-_\Omega f(z)|_\Omega, \quad z \in \Omega\backslash\partial\Omega.$$

PROOF OF (i). See, e.g., Section 2 in [1].

PROOF OF (ii). Since $\partial\Omega = \{\varphi|_\Omega = 0\}$ and $d(\varphi|_\Omega)(\varsigma) \neq 0$ for $\varsigma \in \partial\Omega$, there exists $C_1 > 0$ such that

$$(3.10) \qquad \|f(\varsigma)\|_{\partial\Omega} \leq C_1 \|f(\varsigma) \wedge d\varphi(\varsigma)\|_\Omega, \quad \varsigma \in \partial\Omega,$$

for all continuous differential forms on Ω.

Now we fix some $\xi \in \partial\Omega$. We may assume that $(\partial\rho/\partial z_1)(\xi) \neq 0$. Then there is a $\mathbb{C}^n$-neighborhood U of ξ such that $\partial\rho, dz_2, \ldots, dz_n$ is a basis of $(1,0)$-forms on $\overline{U}$. Let α_j be the functions on $\overline{U}$ with

$$\partial\varphi = \alpha_1 \partial\rho + \sum_{j=2}^{n} \alpha_j dz_j \text{ on } \overline{U}.$$

Then there exists $C_2 > 0$ with

$$(3.11) \qquad \|\partial\varphi(\varsigma) - \alpha_1(\varsigma)\partial\rho(\varsigma)\|_{\mathbb{C}^n} \leq C_2 \|\partial\varphi(\varsigma) \wedge \partial\rho(\varsigma)\|_{\mathbb{C}^n} = C_2 \gamma(\varsigma)$$

for all $\varsigma \in U$. If now f is a continuous differential form of maximal holomorphic degree on Ω, then, since $f \wedge \overline{\partial}\rho|_\Omega = -f \wedge \partial\rho|_\Omega = 0$ by definition of ρ and Ω,

$$f \wedge d\varphi|_\Omega = f \wedge \overline{\partial}\varphi|_\Omega = f \wedge (\overline{\partial}\varphi - \overline{\alpha}_1\overline{\partial}\rho)|_\Omega$$

and therefore, by (3.10) and (3.11),

$$\begin{aligned}
\|f(\varsigma)\|_{\partial\Omega} &\leq C_1 \|f(\varsigma) \wedge d\varphi(\varsigma)\|_\Omega \\
&= C_1 \|f(\varsigma) \wedge (\overline{\partial}\varphi - \overline{\alpha}_1\overline{\partial}\rho)\|_\Omega \\
&\leq C_1 \|f(\varsigma)\|_\Omega \|\partial\varphi(\varsigma) - \alpha_1(\varsigma)\partial\rho(\varsigma)\|_{\mathbb{C}^n} \\
&\leq C_1 C_2 \gamma(\varsigma) \|f(\varsigma)\|_\Omega \\
&\leq C(\gamma(z) + |\varsigma - z|) \|f(\varsigma)\|_\Omega
\end{aligned}$$

for all $\varsigma \in U \cap \partial\Omega$ and $z \in \mathbb{C}^n$. Taking into account that $\dim_\mathbb{R} \partial\Omega = 2n - 2$ and the singularity of $B(z, \varsigma)$ at $\varsigma = z$ is of order $2n - 1$, this implies estimate (3.5).

PROOF OF (iii). Since

$$d_z[f(\varsigma) \wedge B(z, \varsigma)] = df(\varsigma) \wedge B(z, \varsigma) - d_\varsigma[f(\varsigma) \wedge B(z, \varsigma)], \quad z \in \mathbb{C}^n, \; \varsigma \in \Omega\backslash\{z\},$$

(3.6) and (3.7) follow from Stokes' formula.

PROOF OF (iv) (cf. Theorem 2.5 and its proof in [1]). It follows from

Stokes'formula that

$$\int\limits_{\varsigma\in\partial\Omega} f(\varsigma)\wedge \int\limits_{z\in\mathbb{C}^n} B(z,\varsigma)\wedge\varphi(z) = \int\limits_{\varsigma\in\Omega} df(\varsigma)\wedge \int\limits_{z\in\mathbb{C}^n} B(z,\varsigma)\wedge\varphi(z)$$

$$+(-1)^{n+r}\int\limits_{\varsigma\in\Omega} f(\varsigma)\wedge\overline{\partial}_\varsigma \int\limits_{z\in\mathbb{C}^n} B(z,\varsigma)\wedge\varphi(z).$$

On the other hand, by (3.6) and Fubini's theorem the right hand side of (3.8) is equal to

$$\int\limits_{\varsigma\in\partial\Omega} f(\varsigma)\wedge \int\limits_{z\in\mathbb{C}^n} B(z,\varsigma)\wedge\varphi(z) - \int\limits_{\varsigma\in\Omega} df(\varsigma)\wedge \int\limits_{z\in\mathbb{C}^n} B(z,\varsigma)\wedge\varphi(z)$$

$$+(-1)^{n+r}\int\limits_{\varsigma\in\Omega} f(\varsigma)\wedge \int\limits_{z\in\mathbb{C}^n} B(z,\varsigma)\wedge d\varphi(z).$$

Hence, the right hand side of (3.8) is equal to

$$(-1)^{n+r}\int\limits_{\varsigma\in\Omega} f(\varsigma)\wedge\left[\overline{\partial}_\varsigma \int\limits_{z\in\mathbb{C}^n} B(z,\varsigma)\wedge\varphi(z) + \int\limits_{z\in\mathbb{C}^n} B(z,\varsigma)\wedge d\varphi(z)\right].$$

Since f is of bidegree (n,r), we may assume that φ is of bidegree $(0, n-r-1)$. Then, by the Bochner-Martinelli-Koppelman formula, the expression in brackets in the last integral is equal to $\varphi(\varsigma)$. This completes the proof of (3.8). (3.9) follows from (3.8) by Stokes'formula, which can be applied in view of estimates (3.3), (3.5) and relation (3.6). ∎

The jump representation (3.8), resp. (3.9), need not to be closed if f is closed. Therefore now we modify this jump. Doing this we have to solve a $\overline{\partial}$-equation and, since we want to preserve (at least "almost") estimate (3.3), we have to do this with appropriate uniform estimates. For that we need some convexity hypotheses on Ω.

We need the following two definitions from [21] and [7].

DEFINITION 3.2. An open set $D\subset\subset\mathbb{C}^n$ will be called *local q-convex*, $0\leq q\leq n-1$, if there exist a neighborhood U of $\overline{D}$ and a finite number of real C^2-functions $\rho_1,\ldots,\rho_N$ on U such that:

(i) $D = \{\rho_1 < 0\}\cap\cdots\cap\{\rho_N < 0\}$.

(ii) $d\rho_{k_1}(z)\wedge\cdots\wedge d\rho_{k_\ell}(z)\neq 0$ if $1\leq k_1 < \cdots < k_\ell \leq N$ and $z\in U$ with $\rho_{k_1}(z) = \cdots = \rho_{k_\ell}(z) = 0$.

(iii) Let $MO(n, n-q-1)$ be the set of all complex $n\times n$-matrices which define an orthogonal projection from $\mathbb{C}^n$ onto some $(n-q-1)$-dimensional subspace of $\mathbb{C}^n$, and let Δ be the simplex of all $\lambda\in\mathbb{R}^N_+$ with $\lambda_1+\cdots+\lambda_N = 1$. Set

$$\rho_\lambda = \lambda_1 \rho_1 + \cdots + \lambda_N \rho_N \ \text{ for } \lambda \in \Delta.$$

Then, for all $\lambda \in \Delta$ and $z \in U$, the Levi form $L_z^{\mathbb{C}^n}(\rho_\lambda)$ has at least $q+1$ positive eigenvalues, and moreover there exist a C^∞-map $Q: \Delta \to MO(n, n-q-1)$ and constants A, $\alpha > 0$ such that

$$\text{Re}\left[2 \sum_{j=1}^n \frac{\partial \rho_\lambda(\varsigma)}{\partial \varsigma_j}(\varsigma_j - z_j) - \sum_{j,k=1}^n \frac{\partial^2 \rho_\lambda(\varsigma)}{\partial \varsigma_j \partial \varsigma_k}(\varsigma_j - z_j)(\varsigma_k - z_k) \right]$$

$$\geq \rho_\lambda(\varsigma) - \rho_\lambda(z) + \alpha |\varsigma - z|^2 - A|Q(\lambda)(\varsigma - z)|^2$$

for all $\lambda \in \Delta$ and $z, \varsigma \in U$.

DEFINITION 3.3. An open set $D \subset\subset \mathbb{C}^n$ will be called *linearly q-convex*, $0 \leq q \leq n-1$, if there exists a real C^2-function ρ on a neighborhood U of $\overline{D}$ such that:

(i) $D = \{\rho < 0\}$.

(ii) $d\rho(z) \neq 0$ if $z \in \partial D$.

(iii) There exist holomorphic coordinates $h_1, \ldots, h_n$ on U such that ρ is strictly convex (in the linear sense) with respect to the real coordinates $\text{Re}\,h_1$, $\text{Im}\,h_1, \ldots, \text{Re}\,h_{q+1}, \text{Im}\,h_{q+1}$.

REMARK 3.4 (to these definitions).

I. *Local q-convex open sets* in the sense of Definition 3.2 are *local q-convex domains* in the sense of Definition 2.3 in [21], although there the following additional condition is assumed:

(3.10) $\qquad (d\rho_{k_1}(z) - d\rho_{k_2}(z)) \wedge \cdots \wedge (d\rho_{k_1}(z) - d\rho_{k_\ell}(z)) \neq 0$

if $1 \leq k_1 < \cdots < k_\ell \leq N$ and $z \in U$ with $\rho_{k_1}(z) = \cdots = \rho_{k_\ell}(z) = 0$. The point is that, starting with functions $\rho_1, \ldots, \rho_N$ as in Definition 3.2, this additional condition can always be achieved by a C^2-small change of these functions. This follows easily from a lemma of M. Morse, resp. Sard's theorem (cf. Proposition 0.5 in Appendix B of [14]). Recall that condition (3.10) was used in [21] to ensure the smoothness of the manifolds Γ_K introduced in Section 4.2 of [21].

II. In [7] and [8] *linearly q-convex open sets* in the sense of Definition 5.2 are called *local q-convex domains*. We changed their name in order to distinguish them from the local q-convex open sets defined in Definition 3.3.

THEOREM 3.5. *Let* $D \subset\subset \mathbb{C}^n$ *be a linearly q-convex open set*, M *a closed* C^2-*hypersurface in some neighborhood of* $\overline{D}$ *such that the intersection* $M \cap \partial D$ *is transversal. Set*

$$\Omega = M \cap \overline{D} \quad \text{and} \quad \partial\Omega = M \cap \partial D.$$

Then there exists a linear operator

(3.12) $$S: C_{n,r}^0(\Omega) \to C_{n,r}^0(D\backslash\Omega), \quad 0 \le r \le n-1,$$

which has the following properties:

(i) *There is a constant $C > 0$ with*

(3.13) $$\|Sf(z)\| \le C(1 + |\ell n \operatorname{dist}(z,\Omega)|^3) \max_{\varsigma \in \Omega} \|f(\varsigma)\|$$

for all $f \in C_{n,*}^0(\Omega)$ and $z \in D\backslash\Omega$. Moreover, if $f \in C_{n,*}^\alpha(\Omega)$ for some $\alpha > 0$, then Sf admits continuous extensions from both sides to $\Omega\backslash\partial\Omega$.

(ii) *If $f \in C_{n,r}^0(\Omega)$ and $df \in C_{n,r+1}^0(\Omega)$ where $n-q \le r \le n-1$, then*

(3.14) $$dSf + Sdf = 0 \text{ on } D\backslash\Omega.$$

(iii) *If $f \in C_{n,r}^0(\Omega)$ and $df \in C_{n,r+1}^0(\Omega)$ where $n-q-1 \le r \le n-1$, then*

(3.15) $$(-1)^{n+r} \int_\Omega f \wedge \varphi = \int_{D\backslash\Omega} dSf \wedge \varphi + (-1)^{n+r} \int_{D\backslash\Omega} Sf \wedge d\varphi$$

for all C^∞-forms φ with compact support in D. (The integrals on the right hand side of (3.15) exist by estimate (3.13) and relation (3.14).) If, moreover, f is Hölder continuous on Ω and S^+f, resp. S^-f, is the continuous extension of Sf from the left, resp. the right, to $\Omega\backslash\partial\Omega$ (which then exists by part (i) of the theorem), then

(3.16) $$(-1)^{n+r}f(z) = S^+f(z)|_\Omega - S^-f(z)|_\Omega, \quad z \in \Omega\backslash\partial\Omega.$$

PROOF. Denote by $L_*^1(D)$ the space of differential forms with integrable coefficients on D. Let

$$H: L_*^1(D) \cap C_{n,*}(D) \to C_{n,*}^0(D)$$

be the operator constructed in Section 4 of [7] for D (H is a version of the Grauert/Lieb-Henkin integral operator). Then

$$H(L_*^1(D) \cap C_{n,r}^0(D)) \subseteq C_{n,r-1}^0(D), \quad 1 \le r \le n,$$

and it follows from Theorem 2.1 in [8] that

(3.17) $$g = dHg + Hdg$$

if $g \in L^1_*(D) \cap C^0_{n,r}(D)$ and $dg \in L^1_*(D) \cap C^0_{n,r+1}(D)$ with $n - q \leq r \leq n$. Let B_Ω and $B_{\partial\Omega}$ be the operators defined by (3.1) and (3.2). Estimate (3.5) then shows in particular that $B_{\partial\Omega}f$ belongs to $L^1_*(D)$ for all continuous differential forms f on $\partial\Omega$. Hence $HB_{\partial\Omega}$ is a linear operator from $C^0_{n,*}(\Omega)$ to $C^0_{n,*}(D)$ with

$$HB_{\partial\Omega}(C^0_{n,r}(\Omega)) \subseteq C^0_{n,r}(D), \quad 0 \leq r \leq n - 1.$$

Using the complete information of estimate (3.5) it follows from Theorem 2.1 in [8] that there is a constant $C_1 > 0$ with

(3.18) $$\|HB_{\partial\Omega}f(z)\|_{\mathbb{C}^n} \leq C_1(1 + |\ell n \operatorname{dist}(z, \partial\Omega)|^3) \max_{\varsigma \in \partial\Omega} \|f(\varsigma)\|_\Omega$$

for all $f \in C^0_{n,*}(\Omega)$ and $z \in D$. Setting

(3.19) $$S = B_\Omega - HB_{\partial\Omega}$$

now we obtain a linear operator from $C^0_{n,*}(\Omega)$ to $C^0_{n,*}(D\backslash\Omega)$ with

$$S(C^0_{n,r}(\Omega)) \subseteq C^0_{n,r}(D\backslash\Omega), \quad 0 \leq r \leq n - 1.$$

It remains to prove that S has the properties (i)-(iii).

PROOF OF (i). Estimate (3.13) follows from (3.3) and (3.18). That Sf admits continuous extensions from both sides to $\Omega\backslash\partial\Omega$ if f is Hölder continuous, follows from the corresponding property of $B_\Omega f$ (Proposition 3.1 (i)) and the fact that $HB_{\partial\Omega}$ is continuous on D.

PROOF OF (ii). In view of relation (3.7) and estimate (3.5), the forms $B_{\partial\Omega}f$ and $dB_{\partial\Omega}f = B_{\partial\Omega}df$ belong to $L^1_*(D) \cap C^0_{n,*}(D)$. Hence, by (3.17),

(3.20) $$B_{\partial\Omega}f = dHB_{\partial\Omega}f + HdB_{\partial\Omega}f = dHB_{\partial\Omega}f + HB_{\partial\Omega}df$$

and therefore, by (3.6),

$$dSf + Sdf = dB_\Omega f + B_\Omega df - B_{\partial\Omega}f = 0.$$

PROOF OF (iii). The difference between the right hand sides of (3.8) and (3.15) is

$$\int_{D\backslash\Omega} d(HB_{\partial\Omega}f \wedge \varphi).$$

Since $HB_{\partial\Omega}f \wedge \varphi$ has compact support in D and, by (3.20), $d(HB_{\partial\Omega}f \wedge \varphi)$ is continuous in D, this implies that this difference is zero. Hence (3.15) follows from (3.8). (3.16) follows from (3.9).   ∎

I.4 - Local homotopy formulas with Hölder estimates for the tangential Cauchy-Riemann equation

We use the notations from Section 3. The main result of this chapter is a local homotopy formula, which we present in two versions in the following Theorems 4.1 and 4.2.

THEOREM 4.1. *Let M be a real C^2-hypersurface in $\mathbb{C}^n$, $\xi \in M$, and φ a $(q+1)$-convex function on M, $1 \le q \le (n-1)/2$. Further let $\xi \in M$ be a point with $d\varphi(\xi) \neq 0$. Set*

$$\Omega_\tau = \{z \in M : \varphi(z) < \varphi(\xi) \text{ and } |\xi - z| < \tau\} \text{ for } \tau > 0.$$

Then there exists $\delta > 0$ such that, for all α, β with $0 < \alpha < \beta \le \delta$, we can find a continuous linear operator

$$T: C^0_{n,r}(\overline{\Omega}_\beta) \to C^{<1/2}_{n,r-1}(\overline{\Omega}_\alpha), \quad n - q \le r \le n - 1,$$

such that

(4.1) $$f|_{\Omega_\alpha} = dTf - Tdf$$

for all $f \in C^0_{n,r}(\overline{\Omega}_\beta)$ with $df \in C^0_{n,r+1}(\overline{\Omega}_\beta)$ and $n - q \le r \le n - 1$.

PROOF. Set $U_\tau = \{z \in \mathbb{C}^n : |z - \xi| < \tau\}$ for $\tau > 0$. Then, by definition of $(q+1)$-convex functions, we have a number $\delta > 0$ and real C^2-functions ψ, ρ_+, ρ_- on U_δ such that $\psi = \varphi$ on $M \cap U_\delta$, $M \cap U_\delta = \{\rho_\pm = 0\}$, $d\rho_\pm(z) \neq 0$ for all $z \in U_\delta$, $\rho_+ \rho_- < 0$ on $U_\delta \setminus M$, and the forms

$$L^{\mathbb{C}^n}_z(\lambda\psi + (1-\lambda)\rho_\pm), \quad z \in U_\delta, \ 0 \le \lambda \le 1,$$

have at least $q + 1$ positive eigenvalues. Since $d\varphi(\xi) \neq 0$ we may also assume that $d\psi(z) \neq 0$ for all $z \in U_\delta$. Set

$$W_\tau = \{z \in U_\tau : \psi(z) < \varphi(\xi)\} \text{ and } W^\pm_\tau = \{z \in W_\tau : \rho_\pm(z) < 0\}$$

for $0 < \tau \le \delta$. Since $\rho_+ \rho_- < 0$ on $U_\delta \setminus M$, then

$$W_\tau \setminus M = W^+_\tau \cup W^-_\tau \text{ and } W^+_\tau \cap W^-_\tau = \emptyset, \quad \tau > 0.$$

By Lemma 2.4 in [21] and Lemma 3.1 in [7] we may moreover assume that, for $0 < \tau \le \delta$, W^+_τ and W^-_τ are local q-convex, and, for $0 < \alpha < \beta \le \delta$, there exists linearly q-convex open sets between W_α and W_β.

Now let α, β with $0 < \alpha < \beta \le \delta$ be given, and let a linearly q-convex open set D with $W_\alpha \subseteq D \subseteq W_\beta$ be fixed.

For $\gamma > 0$ we denote by $B_{n,r}^\gamma(W_\tau^\pm)$, $0 < \tau \le \delta$, $0 \le r \le n$, the Banach spaces of forms $f \in C_{n,r}^0(W_\tau^\pm)$ with

$$\sup_{z \in W_\tau^\pm} \|f(z)\| \, [\mathrm{dist}(z, M)]^\gamma < \infty.$$

In view of Theorems 4.11 and 4.12 (i) in [21] then there exist linear operators

$$H_\pm : B_{n,r}^\gamma(W_\beta^\pm) \to \cap_{\varepsilon > 0} C_{n,r-1}^{1/2-\gamma-\varepsilon}(\overline{W}_\beta^\pm), \quad 0 \le \gamma < 1/2, \ n - q \le r \le n,$$

such that

(4.2)
$$\left.\begin{array}{l} f = dH_\pm f + H_\pm df \ \text{ on } W_\alpha^\pm \text{ for all } f \in B_{n,r}^\gamma(W_\alpha^\pm) \text{ with} \\[2mm] df \in B_{n,r+1}^\gamma(W_\alpha^\pm), \ 0 \le \gamma < 1/2, \ n - q \le r \le n - 1, \end{array}\right\}$$

and

(4.3)
$$\left.\begin{array}{l} \text{the operators } H_\pm \text{ are bounded from } B_{n,r}^\gamma(W_\alpha^\pm) \\[2mm] \text{to } C_{n,r-1}^{1/2-\gamma-\varepsilon}(\overline{W}_\alpha^\pm), \text{ for } \gamma, \ \varepsilon \text{ with } 0 \le \gamma < 1/2 \\[2mm] \text{and } 0 < \varepsilon < 1/2 - \gamma, \ n - q \le r \le n. \end{array}\right\}$$

Further, let $\Omega := M \cap \overline{D}$ and $\partial\Omega := M \cap \partial D$, and let $S : C_{n,*}^0(\Omega) \to C_{n,*}^0(D\backslash\Omega)$ be the jump operator from Theorem 3.5. By estimate (3.13), S defines linear operators

$$S_\pm : C_{n,r}^0(\overline{\Omega}_\beta) \to \cap_{\varepsilon > 0} B_{n,r}^\varepsilon(W_\alpha^\pm), \quad 0 \le r \le n - 1,$$

which are, for each $\varepsilon > 0$, bounded as operators with values in $B_{n,r}^\varepsilon(W_\alpha^\pm)$. In view of (4.3) then the operators $H_\pm S_\pm$ are bounded as operators from $C_{n,r}^0(\overline{\Omega}_\beta)$ to $C_{n,r-1}^{<1/2}(\overline{W}_\alpha^\pm)$, $n - q \le r \le n - q$. Therefore, setting

(4.4)
$$(-1)^{n+r-1} Tf = (H_+ S_+ f)|_{\overline{\Omega}_\alpha} - (H_- S_- f)|_{\overline{\Omega}_\alpha}$$

for $f \in C_{n,r}^0(\overline{\Omega}_\beta)$, $n - q \le r \le n - 1$, we obtain operators

$$T : C_{n,r}^0(\overline{\Omega}_\beta) \to \cap_{\varepsilon > 0} C_{n,r-1}^{1/2-\varepsilon}(\overline{\Omega}_\alpha), \quad n - q \le r \le n - 1.$$

It remains to prove the homotopy formula (4.1). Since $\Omega_\alpha = M \cap W_\alpha$, for this it is sufficient to fix an orientation on Ω_α and to prove that

(4.5)
$$\int_{\Omega_\alpha} f \wedge \varphi = (-1)^{n+r-1} \int_{\Omega_\alpha} Tf \wedge d\varphi - \int_{\Omega_\alpha} Tdf \wedge \varphi$$

for all C^∞-forms φ with compact support in W_α. From (4.2) and (3.14) it follows that

$$dH_\pm S_\pm f = S_\pm f - H_\pm dS_\pm f = S_\pm f + H_\pm S_\pm df$$

and

$$dH_\pm S_\pm df = -dS_\pm f.$$

If the orientation of Ω_α is induced from W_α^+, this implies by Stokes'formula that

$$(-1)^{n+r-1} \int_{\Omega_\alpha} Tf \wedge d\varphi = \int_{W_\alpha \backslash \Omega_\alpha} Sf \wedge d\varphi + \int_{W_\alpha^+} H_+ S_+ df \wedge d\varphi + \int_{W_\alpha^-} H_- S_- df \wedge d\varphi$$

and

$$-\int_{\Omega_\alpha} Tdf \wedge \varphi = (-1)^{n+r} \int_{W_\alpha \backslash \Omega_\alpha} dSf \wedge \varphi - \int_{W_\alpha^+} H_+ S_+ df \wedge d\varphi - \int_{W_\alpha^-} H_- S_- df \wedge d\varphi$$

for all C^∞-forms φ with compact support in W_α. This implies that the right hand side of (4.5) is equal to

$$\int_{W_\alpha \backslash \Omega_\alpha} Sf \wedge d\varphi + (-1)^{n+r} \int_{W_\alpha \backslash \Omega_\alpha} dSf \wedge \varphi.$$

Hence (4.5) follows from Theorem 3.5 (iii). ∎

THEOREM 4.2. *Let M be a q-convex-concave real C^2-hypersurface in $\mathbb{C}^n$, $1 \le q \le (n-1)/2$, φ a real C^2-function on M, and $\xi \in M$ a point such that $d\varphi(\xi) = 0$ and the Hessian matrix of φ at ξ is positive definite[*]. Set*

$$\Omega_\tau = \{z \in M : \varphi(z) < \varphi(\xi) + \tau\} \text{ for } \tau > 0.$$

Then there exists $\delta_0 > 0$ such that, for all δ with $0 < \delta \le \delta_0$, there exists a continuous linear operator

$$T : C_{n,r}^0(\overline{\Omega}_\delta) \to C_{n,r-1}^{<1/2}(\overline{\Omega}_\delta), \quad n - q \le r \le n - 1,$$

such that

(4.6) $$f = dTf - Tdf \text{ on } \overline{\Omega}_\delta.$$

for all $f \in C_{n,r}^0(\overline{\Omega}_\delta)$ with $df \in C_{n,r+1}^0(\overline{\Omega}_\delta)$ and $n - q \le r \le n - 1$.

PROOF. Since M is q-convex-concave, we can find a $\mathbb{C}^n$-neighborhood U of ξ and real C^2-functions ρ_+, ρ_- on U such that $M \cap U = \{\rho_\pm = 0\}$, $d\rho_\pm(z) \ne 0$ for all $z \in U$, $\rho_+ \rho_- < 0$ on $U \backslash M$, and the forms

$$L_z^{\mathbb{C}^n}(\rho_\pm), \quad z \in U,$$

---

[*] Since $d\varphi(\xi)=0$, this property of the Hessian matrix is independent of the choice of the local coordinates on M.

have at least $q+1$ positive eigenvalues. Since $d\varphi(\xi) = 0$ and the Hessian matrix of φ is positive definite, after shrinking U, moreover we can find a real C^2-extension ψ of φ to U which is strictly convex (with respect to the real-linear structure of $\mathbb{C}^n$) and such that $d\psi(\xi) = 0$. Set

$$D_\delta = \{z \in U : \psi(z) < \varphi(\xi) + \delta\} \text{ and } D_\delta^\pm = \{z \in D_\delta : \rho_\pm(z) < 0\}$$

for $\delta > 0$. Then it is clear that, for sufficiently small $\delta > 0$, the set D_δ is linearly $(n-1)$-convex and the sets D_δ^+ and D_δ^- are local q-convex.

To complete the proof now we proceed in the same way as in the proof of Theorem 4.1. (Instead of the three different sets $W_\alpha \subseteq D \subseteq W_\beta$ in the proof of Theorem 4.1, now we have to consider the one set D_δ.) ∎

I.5 - First global consequences of the local homotopy formula

DEFINITION 5.1. Let M be a real C^2-hypersurface in an n-dimensional complex manifold X, and q an integer with $0 \le q \le (n-1)/2$. An open subset D of M will be called *strictly q-convex in M* if D is relatively compact in M, D is q-convex-concave (as a hypersurface in X) and there exists a $(q+1)$-convex function φ defined in some open M-neighborhood U of the boundary ∂D of D

$$D \cap U = \{z \in U : \varphi(z) < 0\} \text{ and } d\varphi(z) \ne 0 \text{ for all } z \in \partial D.$$

If, in this case, U can be chosen as an open M-neighborhood of $\overline{D}$, then D will be called strictly *completely q-convex in M*.

REMARK. The case $\partial D = \emptyset$ (*i.e.* D is compact) is also possible in this definition, but then D cannot be *completely* q-convex.

THEOREM 5.2. *Let M be a real C^2-hypersurface in an n-dimensional complex manifold X, E a holomorphic vector bundle on X, and D a strictly q-convex open subset in M, $1 \le q \le (n-1)/2$. Then, for all r with $n-q \le r \le n-1$:*

(i) *There exist continuous linear operators*

$$T : C_{n,r}^0(\overline{D}, E) \to C_{n,r-1}^{<1/2}(\overline{D}, E) \text{ and } K : C_{n,r}^0(\overline{D}, E) \to C_{n,r}^{<1/2}(\overline{D}, E)$$

such that

$$f + Kf = dTf - Tdf$$

for all $f \in C_{n,r}^0(\overline{D}, E)$ such that df is also continuous on $\overline{D}$.

(ii) *The space $E_{n,r}^{<1/2}(\overline{D}, E)$ is topologically closed and finite codimensional in $Z_{n,r}^0(\overline{D}, E)$, i.e.*

(5.2) $$\dim H_{n,r}^{<1/2}(\overline{D}, E) < \infty.$$

(iii) *There exists a continuous linear operator* $A: Z_{n,r}^0(\overline{D}, E) \to C_{n,r-1}^{<1/2}(\overline{D}, E)$ *with* $dAf = f$ *for all* $f \in E_{n,r}^{<1/2}(\overline{D}, E)$.

(iv) *The image of the restriction map*

$$\text{germ } Z_{n,n-q-1}^{<1/2}(\overline{D}, E) \to \text{germ } Z_{n,n-q-1}^0(\overline{D}, E)$$

is dense with respect to uniform convergence on $\overline{D}$.

PROOF. Parts (i)-(iii) follow by standard arguments from Theorems 4.1 and 4.2 (see, e.g., the proofs of Lemma 2.3.1 in [14] and Proposition 3 in Appendix 2 in [14]).

To prove part (iv) we consider an open M-neighborhood U of $\overline{D}$ and a form $f \in Z_{n,n-q-1}^0(U, E)$. Choose a strictly q-convex open set Ω in M such that $D \subset\subset \Omega \subset\subset U$. By a theorem of Airapetjan and Henkin (Theorem 7.2.3 in [2]), locally, f can be approximated uniformly by closed $(n, n-q-1)$-forms of class C^1. By part (iii), applied to Ω, and a partition of unity argument, this implies that f can be approximated uniformly on $\overline{\Omega}$ by forms in $Z_{n,n-q-1}^{<1/2}(\overline{\Omega}, E)$. ∎

I.6 - q-convex extension elements

DEFINITION 6.1. Let M be a real C^2-hypersurface in an n-dimensional complex manifold X, and q an integer with $1 \le q \le (n-1)/2$.

(i) An *affine q-convex configuration for* M is an ordered collection $[U, D; \rho, \rho_+, \rho_-]$ where $D \subset\subset U \subset\subset X$ are open sets ($D = \emptyset$ is possible) and ρ, ρ_+, ρ_- are real C^2-functions on U such that the following conditions are fulfilled:

 • M is closed in some neighborhood of $\overline{\overline{U}}$;
 • U is biholomorphically equivalent to the ball in $\mathbb{C}^n$;
 • D is relatively compact in U, $D = \{\rho < 0\}$, $d\rho(z) \neq 0$ for $z \in \partial D$;
 • $M \cap U = \{\rho_+ = 0\} = \{\rho_- = 0\}$, $d\rho_\pm(z) \neq 0$ for all $z \in U$, and $\rho_+ \rho_- < 0$ on $U \backslash M$;
 • $d\rho(z) \wedge d\rho_\pm(z) \neq 0$ for all $z \in M \cap \partial D$;
 • for all $z \in U$ and $\lambda \in [0, 1]$ the forms $L_z^X(\lambda \rho + (1-\lambda)\rho_\pm)$ have at least $q+1$ positive eigenvalues.

(ii) A *q-convex bump for* M is an ordered collection $[U, D_1, D_2; \rho_1, \rho_2, \rho_+, \rho_-]$ such that $[U, D_j; \rho_j, \rho_+, \rho_-]$, $j = 1, 2$, are affine q-convex configurations for M with $D_1 \subseteq D_2$.

(iii) A *q-convex extension element in* M is an ordered couple $[\Theta_1, \Theta_2]$ such that $\Theta_1 \subseteq \Theta_2$ are open subsets with C^2-boundaries in M satisfying the following condition: there exists a q-convex bump $[U, D_1, D_2; \rho_1, \rho_2, \rho_+, \rho_-]$ for M with

$$\Theta_2 = \Theta_1 \cup (M \cap D_2), \quad \Theta_1 \cap D_2 = M \cap D_1, \quad \text{and} \quad (\overline{\Theta_1 \backslash D_2}) \cap (\overline{\Theta_2 \backslash \Theta_1}) = \emptyset.$$

Here $\overline{\Theta}_j$ is the closure of Θ_j in M, $j = 1, 2$.

REMARKS.

I. If $[U, D; \rho, \rho_+, \rho_-]$ is an affine q-convex configuration for M, then it is clear that the restriction of ρ to $M \cap U$ is $(q + 1)$-convex and hence $M \cap D$ is strictly q-convex (Definition 5.1). Also it is clear that then the sets $D_\pm := D \cap \{\rho_\pm < 0\}$ are strictly q-convex C^2-intersections in the sense of Definition 0.1 in [21] and therefore we may apply Theorem 0.2 from [21] to these sets.

II. The following obvious fact will be used permanently: if $[U, D; \rho, \rho_+, \rho_-]$ is an affine q-convex configuration for M, then for any real C^2-function ε with compact support in U which is sufficiently small in the C^2-topology, the collection $[U, D_\varepsilon; \rho + \varepsilon, \rho_+, \rho_-]$ with $D_\varepsilon := \{\rho + \varepsilon < 0\}$ is also an affine q-convex configuration for M. Hence, for each affine q-convex configuration $[U, D; \rho, \rho_+, \rho_-]$ for M, there exists a basis $\mathfrak{U}$ of neighborhoods of $\overline{D}$ and a family of functions ρ_W, $W \in \mathfrak{U}$, such that, for each $W \in \mathfrak{U}$, the collection $[U, W; \rho_W, \rho_+, \rho_-]$ is also an affine q-convex configuration for M.

III. If $[\Theta_1, \Theta_2]$ is a q-convex extension element in M then, by our definition, Θ_1 need not to be strictly q-convex in M (we even not assume that Θ_1 is relatively compact in M). The case when Θ_1 is strictly q-convex in M is of special interest (some results, the most interesting, are true only under this additional hypothesis). It is easy to see that in this case also Θ_2 is strictly q-convex in M.

LEMMA 6.2. *Let M be a real C^2-hypersurface in an n-dimensional complex manifold X, and $[U, D; \rho, \rho_+, \rho_-]$ an affine q-convex configuration for M, $1 \le q \le (n - 1)/2$. Then*

$$(6.1) \qquad H^{n,r}_{<1/2}(M \cap \overline{D}) = 0 \text{ for } n - q \le r \le n - 1.$$

PROOF. Set $\Omega = M \cap \overline{D}$ and $\partial\Omega = M \cap \partial D$, and let $f \in Z^0_{n,r}(\Omega)$ with $n - q \le r \le n - 1$ be given. By definition of affine q-convex configurations, we may assume that U is a ball in $\mathbb{C}^n$, and consider the form $B_{\partial\Omega} f \in C^0_{n,r+1}(\mathbb{C}^n \backslash \partial\Omega)$ defined by (3.2). By Proposition 3.1 this form is closed (see (3.7)) and satisfies estimate (3.5). Therefore it follows from Theorem 1.2 in [8] that there exists $v \in C^0_{n,r}(D)$ with

$$B_{\partial\Omega} f|_D = dv$$

and

$$(6.2) \qquad \sup_{z \in D} \|v(z)\| \, [1 + |\ell n \, \text{dist}(z, \partial\Omega)|]^{-3} < \infty.$$

Let $B_\Omega f \in C^0_{n,r}(\mathbb{C}^n \backslash \Omega)$ be the form defined by (3.1), set $D_\pm = D \cap \{\rho_\pm < 0\}$ and

$$f_\pm = B_\Omega f - v \text{ on } D_\pm.$$

By Proposition 3.1 (see (3.6) the forms $f_\pm$ are closed on $D_\pm$, and it follows from estimate (3.3) in Proposition 3.1 and estimate (6.2) that

$$\sup_{z \in D_\pm} \|f_\pm(z)\| \left[1 + |\ell n \operatorname{dist}(z, \Omega)|\right]^{-3} < \infty.$$

Since the sets $D_\pm$ are strictly q-convex C^2-intersections in the sense of Definition 0.1 in [21] (cf. the remarks following Definition 6.1), therefore it follows from Theorem 0.2 in [21] that the equations $du_\pm = f_\pm$ can be solved on $D_\pm$ with $u_\pm \in C_{n,r-1}^{<1/2}(\overline{D}_\pm)$. Setting

$$u = (-1)^{n+r}(u_+|_\Omega - u_-|_\Omega),$$

now we obtain a form $u \in C_{n,r-1}^{<1/2}(\Omega)$ and it remains to prove that $du = f$, i.e.

(6.3)
$$\int_\Omega u \wedge d\varphi = (-1)^{n+r} \int_\Omega f \wedge \varphi$$

for all C^∞-forms φ with compact support in D (assume that Ω is oriented by D_+). It follows from Stokes' formula and (3.6) in Proposition 3.1 that

$$(-1)^{n+r+1} \int_D v \wedge d\varphi = \int_D dv \wedge \varphi = \int_D B_{\partial\Omega} f \wedge \varphi = \int_{D\backslash\Omega} dB_\Omega f \wedge \varphi$$

and hence

$$\int_\Omega u \wedge d\varphi = (-1)^{n+r} \int_{D_+} du_+ \wedge d\varphi + (-1)^{n+r} \int_{D_-} du_- \wedge d\varphi$$

$$= (-1)^{n+r} \int_{D\backslash\Omega} (B_\Omega f - v) \wedge d\varphi$$

$$= (-1)^{n+r} \int_{D\backslash\Omega} B_\Omega f \wedge d\varphi + \int_{D\backslash\Omega} dB_\Omega f \wedge \varphi$$

for all C^∞-forms φ with compact support in D. This implies (6.3) in view of Proposition 3.1 (iv). ∎

DEFINITION 6.3 (cf. Section 4 in [15]). For each $\beta > 0$ we fix (for the remainder of this chapter) a non-negative real C^∞-function χ_β on $\mathbb{R}$ such that, for all $x \in \mathbb{R}$, $\chi_\beta(x) = \chi_\beta(-x)$, $|x| \leq \chi_\beta(x) \leq |x| + \beta$, $|\chi'_\beta(x)| \leq 1$, $\chi''_\beta(x) \geq 0$, and $\chi_\beta(x) = |x|$ if $|x| \geq \beta/2$, we set

$$\max_\beta(t, s) = \frac{t+s}{2} + \chi_\beta\left(\frac{t-s}{2}\right) \quad \text{for } t, \ s \in \mathbb{R}.$$

We omit the proof of the following simple lemma:

LEMMA 6.4. *Let* φ, ψ *be two real* C^2*-functions defined on some real* C^2*-manifold* X. *Then, for all* $\beta > 0$ *and* $z \in X$, *the following assertions hold:*

(i) $\max(\varphi(z), \psi(z)) \leq \max_{\beta}(\varphi(z), \psi(z)) \leq \max(\varphi(z), \psi(z)) + \beta$.

(ii) $\max_{\beta}(\varphi(z), \psi(z)) = \max(\varphi(z), \psi(z))$ *if* $|\varphi(z) - \psi(z)| \geq \beta$.

(iii) *There is a number* λ_z *with* $0 \leq \lambda_z \leq 1$ *such that*

$$d \max_{\beta}(\varphi(z), \psi(z)) = \lambda_z d\varphi(z) + (1 - \lambda_z)d\psi(z).$$

(iv) *If* X *is a complex manifold and the forms* $L_z^X(\lambda\varphi + (1 - \lambda)\psi)$, $0 \leq \lambda \leq 1$, *have at least* $q + 1$ *positive eigenvalues, then the form* $L_z^X(\max_{\beta}(\varphi, \psi))$ *has at least* $q + 1$ *positive eigenvalues.*

LEMMA 6.5. *Let* M *be a real* C^2*-hypersurface in an* n*-dimensional complex manifold* X, *and* $[U, D; \rho, \rho_+, \rho_-]$ *an affine* q*-convex configuration for* M, $1 \leq q \leq (n - 1)/2$. *Then, any continuous closed* $(n, n - q - 1)$*-form defined in an* M*-neighborhood of* $M \cap \overline{D}$ *can be approximated uniformly on* $M \cap \overline{D}$ *by continuous closed* $(n, n - q - 1)$*-forms defined on* U.

PROOF. It follows from a theorem of Airapetjan and Henkin (Theorem 7.2.3 in [2]) that, locally, any continuous closed $(n, n - q - 1)$-form in an M-neighborhood of $M \cap \overline{D}$ can be approximated uniformly by closed $(n, n - q - 1)$-forms of class C^1. By Lemma 6.2 and a partition of unity argument, this implies that any continuous closed $(n, n - q - 1)$-form in an M-neighborhood of $M \cap \overline{D}$ can be approximated uniformly on some smaller M-neighborhood of $M \cap \overline{D}$ by Hölder continuous closed $(n, n - q - 1)$-forms. Therefore it is sufficient to prove the lemma for Hölder continuous forms.

Let f be a Hölder continuous closed $(n, n - q - 1)$-form defined in some open M-neighborhood Y of $M \cap \overline{D}$.

Then, by Theorem 3.5 (iii)[*], we can find a neighborhood $\Theta \subseteq U$ of $\overline{D}$ and forms $f_\pm \in Z_{n,n-q-1}^0(\overline{\Theta}_\pm)$ where $\Theta_\pm := \Theta \cap \{\rho_\pm < 0\}$ such that $\Theta \cap M \subseteq Y$ and

$$f|_{\Theta \cap M} = f_+|_{\Theta \cap M} - f_-|_{\Theta \cap M}.$$

Now it is sufficient to approximate the forms $f_\pm$.

Set $U_t = \{z \in \mathbb{C}^n : |z| < t\}$, $t > 0$. We may assume that $U = U_1$. Choose $\varepsilon > 0$ sufficiently small, set $D_\varepsilon = \{\rho < \varepsilon\}$ and $D_\varepsilon^\pm = D_\varepsilon \cap \{\rho_\pm < 0\}$. Then $\overline{D}_\varepsilon \subseteq \Theta \cap U_{1-4\varepsilon}$. Choose $\beta > 0$ sufficiently small, set $\psi_\pm = \max_{\beta}(\rho - \varepsilon, \rho_\pm)$ and

---

[*] Actually here we use only a simple corollary of Theorem 3.5, namely the assertion of this theorem without the estimates near $\partial\Omega$. This corollary follows in an elementary way from the classical Andreotti-Grauert theory, because instead of the solution $u = H_{B_{\partial\Omega}}f$ from [7], [8] of the equation $du = B_{\partial\Omega}f$ which we used in the proof of Theorem 3.5, for the proof of this corollary an arbitrary continuous solution on D of this equation is sufficient.

$W_\pm = \{\psi_\pm < 0\}$. Then, by Lemma 6.4, the forms $L_z^{\mathbb{C}^n}(\psi_\pm)$, $z \in U$, have at least $q+1$ positive eigenvalues, $d\psi_\pm(z) \neq 0$ for $z \in \partial W_\pm$, and $D_\pm \subseteq W_\pm \subseteq D_\varepsilon^\pm$. Choose $C > 0$ sufficiently large and set

$$\varphi(z) = C|z|^2 - C(1 - 3\varepsilon)^2, \quad z \in \mathbb{C}^n.$$

Then $\varphi < 0$ on $U_{1-3\varepsilon}$ and $\varphi > \psi_\pm$ on $U_{1-\varepsilon} \backslash U_{1-2\varepsilon}$. Take a number $\alpha > 0$ sufficiently small and set $\eta_\pm = \max_\alpha(\varphi, \psi_\pm)$ on $U_{1-\varepsilon}$ and $\eta_\pm = \varphi$ on $U \backslash U_{1-\varepsilon}$. Then, by Lemma 6.4, the forms $L_z^{\mathbb{C}^n}(\eta_\pm)$, $z \in U$, have at least $q + 1$ positive eigenvalues, $W_\pm = \{\eta_\pm < 0\}$, $d\eta_\pm(z) \neq 0$ for $z \in \partial W_\pm$, and $\eta_\pm = \varphi$ on $U \backslash U_{1-2\varepsilon}$. By a lemma of M. Morse (see, e.g., Proposition 0.5 in Appendix B of [15]) this implies that U is a non-degenerate q-convex extension in the sense of Definition 12.1 in [15] of both W_+ and W_-. Therefore it follows from Theorem 12.11 in [15] that the forms $f_\pm$ can be approximated uniformly on $\overline{W}_\pm$ (and hence on $D_\varepsilon \cap M$) by continuous closed $(n, n - q - 1)$- forms on U.

∎

LEMMA 6.6. *Let M be a real C^2-hypersurface in an n-dimensional complex manifold X, E a holomorphic vector bundle on X, $[\Theta_1, \Theta_2]$ a q-convex extension element in M, $1 \leq q \leq (n - 1)/2$, and $\overline{\Theta}_1$, $\overline{\Theta}_2$ the closures of Θ_1, resp. Θ_2. Then:*

(i) *For all r with $n - q \leq r \leq n - 1$, the restriction map*

(6.4) $$H_{<1/2}^{n,r}(\overline{\Theta}_2, E) \to H_{<1/2}^{n,r}(\overline{\Theta}_1, E)$$

   is surjective.

(ii) *The restriction map*

(6.5) $$\text{germ } H_{<1/2}^{n,r}(\overline{\Theta}_2, E) \to \text{germ } H_{<1/2}^{n,r}(\overline{\Theta}_1, E)$$

   is injective if at least one of the following conditions (a) or (b) is fulfilled:
   (a) $n - q + 1 \leq r \leq n - 1$;
   (b) $r = n - q$ and Θ_1 is strictly q-convex in M.

(iii) *If Θ_1 is strictly q-convex in M, then the restriction map*

   $$\text{germ } Z_{n,n-q-1}^0(\overline{\Theta}_2, E) \to \text{germ } Z_{n,n-q-1}^0(\overline{\Theta}_1, E)$$

   has dense image with respect to uniform convergence on $\overline{\Theta}_1$.

BEGINNING OF THE PROOF. Let $[U, D_1, D_2; \rho_1, \rho_2, \rho_+, \rho_-]$ be a q-convex bump as in Definition 6.1 (iii). Since U is biholomorphically equivalent to the ball, E is holomorphically trivial over U. Since $\overline{(\Theta_1 \backslash D_2)} \cap \overline{(\Theta_2 \backslash \Theta_1)} = \emptyset$, we can find open M-neighborhoods V' and V'' of $\overline{\Theta_2 \backslash \Theta_1}$ such that $V' \subset\subset V'' \subset\subset U$ and

$\overline{V}'' \cap (\overline{\Theta_1 \backslash D_2}) = \emptyset$. Choose a real C^1-function χ on M with $\chi = 1$ on V' and $\chi = 0$ on $M \backslash \overline{V}''$.

PROOF OF (i). Let $f_1 \in Z^0_{n,r}(\overline{\Theta}_1, E)$ with $n - q \le r \le n - 1$ be given. We have to find $u_1 \in C^{<1/2}_{n,r-1}(\overline{\Theta}_1, E)$ and $f_2 \in Z^0_{n,r}(\overline{\Theta}_2, E)$ with $f_2 = f_1 - du_1$ on Θ_1. From Lemma 6.2 we get $u \in C^{<1/2}_{n,r-1}(M \cap \overline{D}_1, E)$ with $f_1 = du$ on $M \cap D_1$. Set

$$u_1 = \begin{cases} 0 & \text{on } \overline{\Theta}_1 \backslash D_2 \\ \chi u & \text{on } M \cap \overline{D}_1 \end{cases} \quad \text{and} \quad f_2 = \begin{cases} f_1 - du_1 & \text{on } \overline{\Theta}_1 \\ 0 & \text{on } \overline{\Theta}_2 \backslash \Theta_1. \end{cases}$$

PROOF OF (ii). Let $\Omega_1 \subseteq \Omega_2$ be open M-neighborhoods of $\overline{\Theta}_1$ and $\overline{\Theta}_2$, respectively, and let $f_2 \in Z^0_{n,r}(\Omega_2, E)$, $u_1 \in C^{<1/2}_{n,r-1}(\Omega_1, E)$ be given such that $du_1 = f_2$ on Ω_1. We have to prove that if one of the conditions (a) or (b) is fulfilled, then there exists an open M-neighborhood $W_2 \subseteq \Omega_2$ of $\overline{\Theta}_2$ with $f_2 \in E^{<1/2}_{n,r}(W_2, E)$.

By Lemma 6.2 and the remarks following Definition 6.1, we can find an open M-neighborhood $D'_2 \subseteq \Omega_2$ of $\overline{D}_2$ and a form $u \in C^{<1/2}_{n,r-1}(D'_2, E)$ with $du = f_2$ on D'_2. Then $(u - u_1) \in Z^{<1/2}_{n,r-1}(\Omega_1 \cap D'_2, E)$.

Now we first assume that condition (a) is fulfilled, i.e. $r - 1 \ge n - q$. Then from Lemma 6.2 we get an open M-neighborhood $D'_1 \subseteq \Omega_1 \cap D'_2$ of $\overline{D}_1$ and a form $v \in C^{<1/2}_{n,r-2}(D'_1, E)$ with $dv = u - u_1$ on D'_1. Choose a real C^1-function ψ with compact support in D'_1 such that $\psi = 1$ in some open M-neighborhood $D''_1 \subset\subset D'_1$ of $\overline{D}_1$. Set $W_2 = \Omega_1 \cup D'_1 \cup D'_2$ and define by

$$u_2 = \begin{cases} u_1 + d(\psi v) & \text{on } D'_1 \cup \Omega_1 \\ u & \text{on } D'_2 \backslash (D'_1 \cup \Omega_1) \end{cases}$$

a form $u_2 \in C^{<1/2}_{n,r-1}(W_2, E)$. Then $du_2 = f_2$ on W_2 and hence $f_2 \in C^{<1/2}_{n,r}(W_2, E)$.

Now we assume that condition (b) is fulfilled. Then, by Lemma 6.5, we find an open M-neighborhood $D'_1 \subseteq \Omega_1 \cap D'_2$ of $\overline{D}_1$ and a sequence $W_k \in Z^{<1/2}_{n,n-q-1}(U, E)$ which converges uniformly on $\overline{D}'_1$ to $u - u_1$. Define by $v_k = (1 - \chi)u_1 + \chi(u - w_k)$ a sequence $v_k \in C^{<1/2}_{n,n-q-1}(\Omega_1 \cup D'_2, E)$. Choose a strictly q-convex open subset W_2 in M with $\Theta_2 \subset\subset W_2 \subset\subset \Omega_1 \cup D'_2$. Then the sequence

$$dv_k = f_2 + d\chi \wedge (u - u_1 - w_k)$$

belongs to $E^{<1/2}_{n,n-q}(\overline{W}_2, E)$ and converges uniformly on $\overline{W}_2$ to f_2. Since, by Theorem 5.1 (ii), $E^{<1/2}_{n,n-q}(\overline{W}_2, E)$ is topologically closed in $Z^0_{n,n-q}(\overline{W}_2, E)$, this implies that $f_2 \in E^{<1/2}_{n,n-q}(\overline{W}_2, E)$. ∎

PROOF OF (iii). Let an open M-neighborhood Ω_1 of $\overline{\Theta}_1$ be given. Take an open neighborhood $W \subset\subset U$ of $\overline{D}_1$ and a function ρ_W such that the collection $[U, W; \rho_W, \rho_+, \rho_-]$ is an affine q-convex configuration for M and $M \cap \overline{W} \subseteq \Omega_1$

(cf. the remarks following Definition 6.1). Further, since Θ_1 and hence Θ_2 is strictly q-convex in M, we can find a strictly q-convex open set Ω_2 in M such that $\overline{\Theta}_2 \subseteq \Omega_2$ and $\overline{\Omega}_2 \subseteq V' \cup W \cup (M\backslash\overline{V}'')$.

Let $f \in Z^0_{n,n-q-1}(\Omega_1, E)$ be given. It is sufficient to find a sequence $f_k \in Z^0_{n,n-q-1}(\Omega_2, E)$ which converges uniformly on $\overline{\Theta}_1$ to f. By Lemma 6.5, there exists a sequence $g_k \in Z^0_{n,n-q-1}(U, E)$ which converges uniformly on $M \cap \overline{W}$ to f. Define by $\tilde{f}_k = (1 - \chi)f + \chi g_k$ a sequence of continuous forms on $\overline{\Omega}_2$. This sequence converges uniformly on $\overline{\Theta}_1$ to f and the sequence $d\tilde{f}_k = d\chi \wedge (f - g_k)$ converges uniformly on $\overline{\Omega}_2$ to zero. Since, by Theorem 5.1, the space $Z^0_{n,n-q}(\overline{\Omega}_2, E) \cap dC^0_{n,n-q-1}(\overline{\Omega}_2, E)$ is topologically closed in $Z^0_{n,n-q}(\overline{\Omega}_2, E)$, we can find a sequence $u_k \in C^0_{n,n-q-1}(\overline{\Omega}, E)$ which also converges uniformly on $\overline{\Omega}_2$ to zero such that $du_k = d\tilde{f}_k$. Set $f_k = \tilde{f}_k - u_k$. ∎

I.7 - q-convex extensions

DEFINITION 7.1. Let M be a real C^2-hypersurface in an n-dimensional complex manifold X, and q an integer with $1 \le q \le (n-1)/2$. In the following definitions, for subsets W of M, we denote by $\overline{W}$ the closure of W in M, and by ∂W the boundary of W in M.

(i) A *q-convex extension in M* is an ordered couple $[D, \Omega]$ of open subsets $D \subseteq \Omega$ of M satisfying the following condition:
∂D is compact and there exist an open M- neighborhood $U_{\partial D}$ of ∂D and a $(q + 1)$-convex function φ defined on $U_\varphi := U_{\partial D} \cup (\Omega \backslash D)$ such that, for some $c_0, c_\infty \in \mathbb{R} \cup \{\infty\}$ with $c_0 < c_\infty$,
(a) $D \cap U_\varphi = \{\varphi < c_0\}$ and $d\varphi(z) \neq 0$ for $z \in \partial D$;
and
(b) the sets $(\Omega \backslash D) \cap \{\varphi \le c\}$, $c_0 \le c < c_\infty$, are compact.

(ii) A *strictly q-convex extension in M* is an ordered couple $[D, \Omega]$ of open subsets $D \subseteq \Omega$ of M such that $\overline{\Omega}\backslash D$ is compact, and there exists a $(q+1)$-convex function φ defined in some open M-neighborhood U_φ of $\overline{\Omega}\backslash D$ such that, for some numbers $c_0 < c_1$,

$$D \cap U_\varphi = \{\varphi < c_0\}, \quad \Omega \cap U_\varphi = \{\varphi < c_1\}$$

and $d\varphi(z) \neq 0$ for $z \in \partial D \cup \partial\Omega$.
If, moreover, this function φ can be chosen so that $d\varphi(z) \neq 0$ also for all $z \in \Omega \backslash \overline{D}$, then $[D, \Omega]$ will be called *non-critical*.

(iii) M will be called *q-convex* if there exists a strictly q-convex open set D in M (Definition 5.1) such that $[D, M]$ is a q-convex extension in M. M will be called *completely q-convex* if $[\emptyset, M]$ is a q-convex extension in M.

REMARK. In the definitions of a (strictly) q-convex extension $[D, \Omega]$ in M we do not assume that D is relatively compact in M.

We want to prove that the local extension properties of the tangential Cauchy-Riemann equation stated in Lemma 6.6 with respect to q-convex extension elements yield corresponding global extension properties with respect to q-convex extensions. The main step in this proof is Theorem 7.10 below which says that any strictly q-convex extension can be obtained by a finite number of q-convex extension elements. This fact corresponds to Lemma 12.3 in [15], however, now the proof is more complicated, since the geometrical picture at critical points of q-convex functions on hypersurfaces is more complicated than for such functions on complex manifolds.

LEMMA 7.2. *Under the hypotheses and with the notations from Theorem 4.2, for all sufficiently small* $\delta > 0$, *there exists an affine q-convex configuration* $[U, D; \rho, \rho_+, \rho_-]$ *for* M *with* $\Omega_\delta = M \cap D$.

PROOF. See the proof of Theorem 4.2.   ∎

LEMMA 7.3. *Let* X *be a complex manifold,* φ_1, φ_2, ψ *real C^2-functions on* X, $\xi \in X$, $\lambda \in [0,1]$, *and* T *a subspace of* $T_\xi^{1,0}(X)$ *such that the forms*

$$L_\xi^X(\lambda \varphi_j + (1 - \lambda)\psi) \quad (j = 1, 2)$$

are positive definite on T. *Then, for each* $\beta > 0$, *also the form*

$$L_\xi^X(\lambda \max_\beta(\varphi_1, \varphi_2) + (1 - \lambda)\psi)$$

is positive definite on T *(see Definition 6.3 for* $\max_\beta$*).*

PROOF. Choose a one dimensional complex submanifold Z of X with $\xi \in Z$ and $T_\xi^{1,0}(Z) \subseteq T$, take a local holomorphic coordinate z in a Z-neighborhood of ξ, and set $x_1 = \operatorname{Re} z$, $x_2 = \operatorname{Im} z$, and $\Delta = \partial^2/\partial x_1^2 + \partial^2/\partial x_2^2$. Then, by hypothesis,

$$(7.1) \qquad (\lambda \Delta \varphi_j + (1 - \lambda)\Delta\psi)(\xi) > 0 \quad (j = 1, 2).$$

We have to prove that, for all $\beta > 0$,

$$(7.2) \qquad (\lambda \Delta \max_\beta(\varphi_1, \varphi_2) + (1 - \lambda)\Delta\psi)(\xi) > 0.$$

Set

$$\alpha = \frac{1}{2} + \frac{1}{2}\chi_\beta'\left(\frac{\varphi_1(\xi) - \varphi_2(\xi)}{2}\right)$$

(see Definition 6.3 for χ_β). Since $|\chi_\beta'| \leq 1$, we see that $0 \leq \alpha \leq 1$. A computation gives

$$(7.3) \qquad \frac{\partial}{\partial t_1} \max_\beta(t_1, t_2)|_{t_j = \varphi_j(\xi)} = \alpha,$$

(7.4)
$$\frac{\partial}{\partial t_2} \max_\beta (t_1, t_2)|_{t_j = \varphi_j(\xi)} = 1 - \alpha,$$

(7.5)
$$\left(\frac{\partial^2}{\partial t_\nu \partial t_\mu} \max_\beta (t_1, t_2)\right)_{\nu, \mu = 1,2} = \frac{1}{4} \chi_\beta'' \left(\frac{t_1 - t_2}{2}\right) \begin{pmatrix} 1 & -1 \\ 1 & 1 \end{pmatrix}.$$

Since $\chi'' \geq 0$, it follows from (7.5) that the matrix on the left hand side of (7.5) is positive semidefinite. Together with (7.3) and (7.4) this implies that

$$(\Delta \max_\beta(\varphi_1, \varphi_2))(\xi) \geq \alpha \Delta \varphi_1 + (1 - \alpha) \Delta \varphi_2.$$

Hence the left hand side of (7.2) is bounded from below by

$$\alpha[\lambda \Delta \varphi_1 + (1 - \lambda) \Delta \psi](\xi) + (1 - \alpha)[\lambda \Delta \varphi_2 + (1 - \lambda) \Delta \psi](\xi).$$

This implies (7.2), in view of (7.1) and since $0 \leq \alpha \leq 1$. ∎

LEMMA 7.4. *Let M be a real C^2-hypersurface in $\mathbb{C}^n$, φ a $(q + 1)$-convex function on M, and $\xi \in M$ a point with $d\varphi(\xi) \neq 0$.*

By definition of $(q+1)$-convex functions, then we have an open ball U in $\mathbb{C}^n$ centered at ξ and real C^2-functions $\tilde{\varphi}$, ρ_+, ρ_- on U with the properties listed in condition (i) *of Proposition 2.3. Since $d\varphi(\xi) \neq 0$ we may moreover assume that $d\tilde{\varphi}(z) \wedge d\rho_+(z) \neq 0$ for all $z \in U$. Set*

$$\rho_{\delta, \beta}(z) = \max_\beta(\tilde{\varphi}(z) - \varphi(\xi), |z - \xi|^2 - \delta^2), \quad z \in U,$$

and $D_{\delta, \beta} = \{\rho_{\delta, \beta} < 0\}$ for δ, $\beta > 0$.

Then there exists $\delta_0 > 0$ such that, for $0 < \delta \leq \delta_0$, there exists $\beta_\delta > 0$ such that, for $0 < \beta \leq \beta_\delta$, the collection $[U, D_{\delta, \beta}; \rho_{\delta, \beta}, \rho_+, \rho_-]$ is an affine q-convex configuration for M.

PROOF. Choose $\delta_0 > 0$ so small that, for $0 < \delta \leq \delta_0$, the three surfaces $\{|z - \xi| = \delta\}$, M and $\{\tilde{\varphi} = \varphi(\xi)\}$ meet transversally. Then the assertion follows easily from Lemmas 6.4 and 7.3. ∎

LEMMA 7.5. *Let M be a real C^2-hypersurface in an n-dimensional complex manifold X, and φ a $(q + 1)$-convex function on M such that the set $\{\varphi = 0\}$ is compact and $d\varphi(z) \neq 0$ for $\varphi(z) = 0$. Then we can find a finite number of open sets $U_1, \ldots, U_N$ of M such that $\{\varphi = 0\} \subseteq U_1 \cup \cdots \cup U_N$ and the following condition is fulfilled:*

Ex$(\varphi; U_1, \ldots, U_N)$. *If $\chi_1, \ldots, \chi_N$ are non-negative real C^2-functions with compact supports in $U_1, \ldots, U_N$, respectively, which are sufficiently small in the C^2-topology, then the couples $[\Theta_j, \Theta_{j+1}]$, $0 \leq j \leq N - 1$, defined by*

$$\Theta_j = \left\{ \varphi < \sum_{\nu=1}^j \chi_\nu \right\}, \quad 0 \leq j \leq N,$$

are q-convex extension elements in M.

PROOF. This follows from Lemma 7.4 and the remarks following Definition 6.1.   ∎

LEMMA 7.6. *Let M be a real C^2-hypersurface in an n-dimensional complex manifold X, and $[D, \Omega]$ a strictly q-convex extension in M. Suppose additional that $[D, \Omega]$ is non critical *[*]*. Then there exists a finite number of sets $\Theta_1, \ldots, \Theta_N$ such that $\Theta_1 = D$, $\Omega = \Theta_N$ and the couples $[\Theta_j, \Theta_{j+1}]$, $1 \leq j \leq N - 1$, are q-convex extension elements in M.*

PROOF. Let U_φ, φ, c_0, c_1 be as in Definition 7.1 (ii) where $d\varphi(z) \neq 0$ for all $z \in \overline{\Omega} \backslash D$. Since the interval $[c_0, c_1]$ is compact, then it is sufficient, for each $c \in [c_0, c_1]$, to find a number $\varepsilon > 0$ such that, for $c - \varepsilon \leq \alpha < \beta \leq c + \varepsilon$, there exists a finite number of sets $\Theta_0, \ldots, \Theta_N$ such that $\Theta_0 = \{\varphi < \alpha\}$, $\Theta_N = \{\varphi < \beta\}$ and the couples $[\Theta_j, \Theta_{j+1}]$, $0 \leq j \leq N - 1$, are q-convex extension elements in M.

Let $c \in [c_0, c_1]$ be given. By Lemma 7.5 we can find a finite number of open subsets $U_1, \ldots, U_N$ of M such that $\{\varphi = c\} \subseteq U_1 \cup \cdots \cup U_N$ and the condition $\mathrm{Ex}(\varphi - c; U_1, \ldots, U_N)$ is fulfilled. Choose non-negative real C^2-functions $\chi_1, \ldots, \chi_N$ with compact supports in $U_1, \ldots, U_N$, respectively, such that $\chi_1 + \cdots + \chi_N = 1$ in some neighborhood of $\{\varphi = c\}$. Since condition $\mathrm{Ex}(\varphi - c; U_1, \ldots, U_N)$ is fulfilled, then we can find $\varepsilon > 0$ such that, for $c - \varepsilon \leq \alpha < \beta \leq c + \varepsilon$, the couples $[\Theta_j, \Theta_{j+1}]$, $0 \leq j \leq N - 1$, defined by

$$\Theta_j = \left\{ \varphi - c < (\alpha - c) \sum_{\nu=1}^{N-j} \chi_\nu + (\beta - c) \sum_{\nu=N-j+1}^{N} \chi_\nu \right\}$$

are q-convex extension elements in M with $\Theta_0 = \{\varphi < \alpha\}$ and $\Theta_N = \{\varphi < \beta\}$.   ∎

In the formulation of the following lemma we want to use certain special notations:

NOTATION 7.7. If χ is a real C^2-function

$$\|\chi\|_2 = \max_{\substack{z \in \mathbb{C}^n \\ 0 \leq |\alpha| \leq 2}} \left| \frac{\partial^{|\alpha|} \chi}{\partial x_1^{\alpha_1} \cdots \partial x_{2n}^{\alpha_{2n}}} (z) \right|.$$

(Here $x_1, \ldots, x_{2n}$ are the real coordinates on $\mathbb{C}^n$.)

For $\xi \in \mathbb{C}^n$ and $\tau > 0$, we denote by $B_\tau(\xi)$ the open ball in $\mathbb{C}^n$ centered at ξ and of radius τ.

Now let $\xi \in \mathbb{C}^n$ and $\tau, \varepsilon > 0$. Then we denote by $BF(\xi, \varepsilon, \tau)$ the set of real C^2-function χ on $\mathbb{C}^n$ which have the following properties:

---

[*] Actually the conclusion of this lemma is true without this additional condition (see Theorem 7.10 below).

- χ has compact support in $B_\tau(\xi)$,

- χ is non-negative,

- χ is positive and constant on $B_{\tau/2}(\xi)$,

- $\|\chi\|_2 \le \varepsilon$.

LEMMA 7.8. *Let M be a real C^2-hypersurface in $\mathbb{C}^n$, φ a $(q+1)$-convex function on M, $1 \le q \le (n-1)/2$, and $\xi \in M$ a non-degenerated critical point of φ which is not the point of a local minimum of φ (i.e. $d\varphi(\xi) = 0$ and the Hessian matrix of φ at ξ is invertible but not positive definite).*

By definition of $(q+1)$-convex functions, then we have real C^2-functions $\tilde{\varphi}$, ρ_+, ρ_- in some $\mathbb{C}^n$-neighborhood U of ξ with the properties listed in condition (i) of Proposition 2.3. Moreover we may assume that $d\tilde{\varphi}(\xi) \ne 0$.

Then, for all sufficiently small $\delta_0 > 0$ with $B_{\delta_0}(\xi) \subseteq U$ and any δ with $0 < \delta < \delta_0$, there exists $\varepsilon > 0$ and τ with $0 < \tau < \delta$ such that the following holds: for all χ_1, $\chi_2 \in BF(\xi; \varepsilon, \tau)$ and any β with $0 < \beta \le \varepsilon$, the collection $[B_{\delta_0}(\xi), D_1, D_2; \rho_1, \rho_2, \rho_+, \rho_-]$ defined by

$$\rho_j(z) = \max_\beta(\tilde{\varphi}(z) - \varphi(\xi), |z - \xi|^2 - \delta^2) + (-1)^{j+1}\chi_j(z), \quad z \in B_{\delta_0}(\xi),$$

and $D_j = \{\rho_j < 0\}$, $j = 1$, 2, is a q-convex bump for M.

PROOF. Set $\tilde{M} = \{\tilde{\varphi} = \varphi(\xi)\}$. Since $d\tilde{\varphi}(\xi) \ne 0$, $d\rho_\pm(\xi) \ne 0$, $M \cap U = \{\rho_\pm = 0\}$, $\tilde{\varphi} = \varphi$ on $M \cap U$, and ξ is a non-degenerate critical point of φ which is neither the point of a local minimum of φ nor the point of a local maximum of φ ($(q+1)$-convex functions do not have local maxima), we can choose $\delta_0 > 0$ so small that $\overline{B_{\delta_0}(\xi)} \subseteq U$, M is closed in some neighborhood of $\overline{B_{\delta_0}(\xi)}$,

(7.6) $d\tilde{\varphi}(z) \ne 0$ for all $z \in \overline{B_{\delta_0}(\xi)}$

(7.7) $d\tilde{\varphi}(z) \wedge d_z|z - \xi|^2 \ne 0$ if $z \in \tilde{M} \cap \overline{B_{\delta_0}(\xi)} \backslash \{\xi\}$

(7.8) $d\tilde{\varphi}(z) \wedge d\rho_\pm(z) \ne 0$ if $z \in M \cap \overline{B_{\delta_0}(\xi)} \backslash \{\xi\}$

(7.9) $d\tilde{\varphi}(z) \wedge d_z|z - \xi|^2 \wedge d\rho_\pm(z) \ne 0$ if $z \in \tilde{M} \cap M \cap \overline{B_{\delta_0}(\xi)} \backslash \{\xi\}$

and

(7.10) $d_z|z - \xi|^2 \wedge d\rho_\pm(z) \ne 0$ if $z \in \overline{B_{\delta_0}(\xi)} \backslash \{\xi\}$.

Let some δ with $0 < \delta < \delta_0$ be given. For $\beta > 0$ we consider the function

$$\gamma_\beta(z) := \max_\beta(\tilde{\varphi}(z) - \varphi(\xi), |z - \xi|^2 - \delta^2), \quad z \in B_{\delta_0}(\xi).$$

By Lemma 7.3, for all $\beta > 0$,

(7.11)
$$\left.\begin{array}{l} \text{the forms } L_z^{\mathbb{C}^n}(\lambda\gamma_\beta + (1-\lambda)\rho_\pm), \ z \in B_{\delta_0}(\xi), \ 0 \le \lambda \le 1, \\ \text{have at least } q+1 \text{ positive eigenvalues.} \end{array}\right\}$$

By (7.7), (7.9) and Lemma 6.4 (iii), we can find a $\mathbb{C}^n$-neighborhood $W_0 \subseteq B_{\delta_0}(\xi)$ of $\tilde{M} \cap \partial B_\delta(\xi)$ such that, for all $\beta > 0$,

(7.12)
$$d\gamma_\beta(z) \ne 0 \text{ if } z \in W_0$$

and

(7.13)
$$d\gamma_\beta(z) \wedge d\rho_\pm(z) \ne 0 \text{ if } z \in M \cap W_0.$$

Set
$$\Gamma = \{z \in B_{\delta_0}(\xi): \max(\tilde{\varphi}(z) - \varphi(\xi), |z - \xi|^2 - \delta^2) = 0\}$$

and choose compact sets $\Gamma_1 \subseteq \tilde{M} \cap B_\delta(\xi)$ and $\Gamma_2 \subseteq \partial B_\delta(\xi) \cap \{\tilde{\varphi} < \varphi(\xi)\}$ such that $\Gamma \subseteq \Gamma_1 \cup \Gamma_2 \cup W_0$. By Lemma 6.4 (i), (ii) then we can find $\varepsilon_1 > 0$ and neighborhoods $W_1 \subseteq B_{\delta_0}(\xi)$ of Γ_1 and $W_2 \subseteq B_{\delta_0}(\xi)\setminus\{\xi\}$ of Γ_2 such that, for $0 < \beta \le \varepsilon_1$,

(7.14)
$$\{\gamma_\beta = 0\} \subseteq \Gamma_1 \cup \Gamma_2 \cup W_0,$$

$\gamma_\beta(z) = \tilde{\varphi}(z) - \varphi(\xi)$ if $z \in W_1$ and $\gamma_\beta(z) = |z - \xi|^2 - \delta^2$ if $z \in W_2$. Since $\xi \notin W_2$ and (7.6), (7.8), (7.10) hold, then $d\gamma_\beta(z) \ne 0$ if $z \in \Gamma_1 \cup \Gamma_2$ and $d\gamma_\beta(z) \wedge d\rho_\pm(z) \ne 0$ if $z \in M \cap (\Gamma_1 \cup \Gamma_2)$. Together with (7.12), (7.13) and (7.14) this implies that

(7.15)
$$d\gamma_\beta(z) \ne 0 \text{ if } z \in \{\gamma_\beta = 0\}$$

and

(7.16)
$$d\gamma_\beta(z) \wedge d\rho_\pm(z) \ne 0 \text{ if } z \in M \cap \{\gamma_\beta = 0\}.$$

Now we choose τ with $0 < \tau < \delta$ so small that

$$\tilde{\varphi} - \varphi(\xi) > \tau^2 - \delta^2 \text{ on } B_\tau(\xi).$$

Then, by Lemma 6.4 (ii), we can find $\varepsilon_2 > 0$ such that,

(7.17)
$$\gamma_\beta = \tilde{\varphi} - \varphi(\xi) \text{ on } B_\tau(\xi) \text{ if } 0 < \beta \le \varepsilon_2.$$

By (7.6)-(7.8) we can find $\varepsilon_3 > 0$ such that, for all real C^2-functions χ on $\mathbb{C}^n$ with $\|\chi\|_2 \le \varepsilon_3$,

(7.18)
$$d(\tilde{\varphi} + \chi)(z) \ne 0 \text{ for all } z \in B_{\delta_0}(\xi)$$

$$(7.19) \qquad d(\tilde{\varphi} + \chi)(z) \wedge d_z |z - \xi|^2 \neq 0 \ \text{ if } \ z \in \tilde{M} \cap (B_{\delta_0}(\xi) \setminus B_{\tau/2}(\xi))$$

and

$$(7.20) \qquad d(\tilde{\varphi} + \chi)(z) \wedge d\rho_{\pm}(z) \neq 0 \ \text{ if } \ z \in M \cap (B_{\delta_0}(\xi) \setminus B_{\tau/2}(\xi)).$$

Since the forms $L_z^{\mathbb{C}^n}(\lambda \tilde{\varphi} + (1 - \lambda)\rho_{\pm})$, $z \in U$, $0 \le \lambda \le 1$, have at least $q + 1$ positive eigenvalues, we can find $\varepsilon_4 > 0$ such that, for all real C^2-functions χ on $\mathbb{C}^n$ with $\|\chi\|_2 \le \varepsilon_4$,

$$(7.21) \qquad \left.\begin{array}{l} \text{the forms } L_z^{\mathbb{C}^n}(\lambda(\tilde{\varphi} + \chi) + (1 - \lambda)\rho_{\pm}), \ z \in B_{\delta_0}(\xi), \ 0 \le \lambda \le 1, \\[2mm] \text{have at least } q + 1 \text{ positive eigenvalues.} \end{array}\right\}$$

Set $\varepsilon = \min(\varepsilon_1, \varepsilon_2, \varepsilon_3, \varepsilon_4)$, let χ_1, $\chi_2 \in BF(\xi; \varepsilon, \tau)$, $0 < \beta \le \varepsilon$, and $\rho_j := \gamma_\beta + (-1)^{j+1}\chi_j$, $D_j := \{\rho_j < 0\}$ $(j = 1, 2)$. We have to prove that $[B_{\delta_0}(\xi), D_1, D_2; \rho_1, \rho_2, \rho_+, \rho_-]$ is a q-convex bump for M. Since χ_1, χ_2 are non-negative, it is clear that $D_1 \subseteq D_2$. It remains to prove that, for $j = 1$, 2, $[B_{\delta_0}(\xi), D_j; \rho_j, \rho_+, \rho_-]$ is an affine q-convex configuration for M, *i.e.* we have to prove that

$$(7.22) \qquad \left.\begin{array}{l} \text{the forms } L_z^{\mathbb{C}^n}(\lambda \rho_j + (1 - \lambda)\rho_{\pm}), \ z \in B_{\delta_0}(\xi), \ 0 \le \lambda \le 1, \\[2mm] \text{have at least } q + 1 \text{ positive eigenvalues;} \end{array}\right\}$$

$$(7.23) \qquad\qquad\qquad D_j \subset\subset B_{\delta_0}(\xi);$$

$$(7.24) \qquad\qquad\qquad d\rho_j(z) \neq 0 \ \text{ if } \ z \in \partial D_j;$$

and

$$(7.25) \qquad\qquad\qquad d\rho_j(z) \wedge d\rho_{\pm}(z) \neq 0 \ \text{ if } \ z \in M \cap \partial D_j.$$

(The other properties listed in Definition 6.1 (i) are clear from the choices of δ_0 and $\rho_{\pm}$.)

(7.22) follows immediately from (7.21).

PROOF OF (7.23). Since $\chi_j = 0$ outside $B_\tau(\xi)$, it follows from Lemma 6.4 (i) that $\rho_j(z) \ge |z - \xi|^2 - \delta^2 \ge 0$ for $z \in B_{\delta_0}(\xi) \setminus B_\delta(\xi)$. Hence $D_j \subseteq B_\delta(\xi) \subset\subset B_{\delta_0}(\xi)$.

PROOF OF (7.24). Since $\rho_j = \gamma_\beta$ outside a compact subset of $B_\tau(\xi)$, it follows from (7.15) that (7.24) holds for $z \in (B_{\delta_0}(\xi) \setminus B_\tau(\xi)) \cap \partial D_j$. Since, by (7.17), $\rho_j = \tilde{\varphi} - \varphi(\xi) + (-1)^{j+1}\chi_j$ on $B_\tau(\xi)$, it follows from (7.18) that (7.24) holds also for $z \in B_\tau(\xi)$.

PROOF OF (7.25). First consider $z \in M \cap \partial D_j \cap (B_{\delta_0}(\xi) \backslash B_\tau(\xi))$. Since $\rho_j = \gamma_\beta$ outside a compact subset of $B_\tau(\xi)$, then it follows from (7.16) that $d\rho_j(z) \wedge d\rho_\pm(z) \neq 0$.

Now let $z \in M \cap \partial D_j \cap (B_\tau(\xi) \backslash B_{\tau/2}(\xi))$. Since $\rho_j = \check{\varphi} - \varphi(\xi) + (-1)^{j+1} \chi_j$ on $B_\tau(\xi)$, then it follows from (7.20) that $d\rho_j(z) \wedge d\rho_\pm(z) \neq 0$.

Finally let $z \in M \cap \partial D_j \cap B_{\tau/2}(\xi)$. Since $\rho_j = \check{\varphi} - \varphi(\xi) + (-1)^{j+1} \chi_j$ and χ_j is a positive constant on $B_{\tau/2}(\xi)$, then $z \neq \xi$ and it follows from (7.8) that $d\rho_j(z) \wedge d\rho_\pm(z) = d\check{\varphi}(z) \wedge d\rho_\pm(z) \neq 0$. ∎

COROLLARY 7.9 (to Lemma 7.8). *Let M be a real C^2-hypersurface in an n-dimensional complex manifold X, φ a $(q+1)$-convex function on M, and $\xi \in M$. Set $D_\tau = \{\varphi < \tau\}$ and suppose that, for some $\alpha < \beta$, the following conditions are fulfilled:*

• *The set $D_\beta \backslash \overline{D}_\alpha$ is relatively compact in M;*

• *$d\varphi(z) \neq 0$ for $z \in M \backslash \{\xi\}$;*

• *$\alpha < \varphi(\xi) < \beta$ and ξ is a non-degenerate critical point of φ which is not the point of a local minimum of φ.*

Then there exists an open M-neighborhood $W \subset\subset D_\beta \backslash \overline{D}_\alpha$ of ξ such that $[D_{\varphi(\xi)} \backslash \overline{W}, D_{\varphi(\xi)} \cup W]$ is a q-convex extension element in M and $[D_\alpha, D_{\varphi(\xi)} \backslash \overline{W}]$, $[D_{\varphi(\xi)} \cup W, D_\beta]$ are non-critical strictly q-convex extensions in M.

THEOREM 7.10. *Let M be a real C^2-hypersurface in an n-dimensional complex manifold X, and $[D, \Omega]$ a strictly q-convex extension in M. Then there exists a finite number of sets $\Theta_1, \ldots, \Theta_N$ such that $\Theta_1 = D$, $\Omega = \Theta_N$ and the couples $[\Theta_j, \Theta_{j+1}]$, $1 \leq j \leq N-1$ are q-convex extension elements in M.*

PROOF. Let U_φ, φ, c_0, c_1 be as in Definition 7.1 (ii). Set $D_\alpha = \{\varphi < \alpha\}$, $\alpha \in \mathbb{R}$. By a lemma of M. Morse (see, e.g., Proposition 0.5 in Appendix B of [15]) we may assume that all critical points of φ are non-degenerate and that, for all $\alpha \in \mathbb{R}$, at most one critical point of φ belongs to ∂D_α.

DEFINITION. For $\alpha < \beta$ we say that D_β is a *finite q-convex extension* of D_α if there is a finite number of sets $\Theta_1, \ldots, \Theta_N$ with $\Theta_1 = D_\alpha$, $\Theta_N = D_\beta$ such that the couples $[\Theta_j, \Theta_{j+1}]$, $1 \leq j \leq N-1$, are q-convex extension elements in M.

We have to prove that D_{c_1} is a finite q-convex extension of D_{c_0}. If $c_0 \leq \alpha < \beta \leq c_1$ such that $\overline{D}_\beta \backslash D_\alpha$ contains no critical point of φ, then it follows from Lemma 7.6 that D_β is a finite q-convex extension of D_α. Therefore it remains to prove the following two assertions:

(i) If $\xi \in \Omega \backslash \overline{D}$ is a critical point of φ which is not the point of a local minimum of φ, then there exists $\varepsilon > 0$ such that $D_{\varphi(\xi)+\varepsilon}$ is a finite q-convex extension of $D_{\varphi(\xi)-\varepsilon}$.

(ii) If $\xi \in \Omega \backslash \overline{D}$ is the point of a local minimum of φ, then there exists $\varepsilon > 0$ such that $D_{\varphi(\xi)+\varepsilon}$ is a finite q-convex extension of $D_{\varphi(\xi)}$.

PROOF OF (i). This follows immediately from Corollary 7.9 and Lemma 7.6.

PROOF OF (ii). Let $\varepsilon > 0$ be sufficiently small such that $U_\varepsilon := \{z \in \Omega \setminus \overline{D}_{\varphi(\xi)} : \varphi(z) < \varphi(\xi) + \varepsilon\}$ has at least two commuted components and let W_ε be the connected component of U_ε containing ξ.

Since ξ is non-degenerate as a critical point of φ and therefore the point of a *strong* local minimum of φ, then $\xi \notin \overline{D}_{\varphi(\xi)}$ and it follows from Lemma 7.2 that $[D_{\varphi(\xi)+\varepsilon} \setminus W_\varepsilon, D_{\varphi(\xi)+\varepsilon}]$ is a q-convex extension element in M. Moreover then it is clear that for sufficiently small $[D_{\varphi(\xi)}, D_{\varphi(\xi)+\varepsilon} \setminus W_\varepsilon]$ is a non critical strictly q-convex extension in M and hence, by Lemma 7.6, $D_{\varphi(\xi)+\varepsilon} \setminus W_\varepsilon$ is a finite q-convex extension of $D_{\varphi(\xi)}$. Together this implies that $D_{\varphi(\xi)+\varepsilon}$ is a finite q-convex extension of $D_{\varphi(\xi)}$. ∎

I.8 - Conclusions

THEOREM 8.1. *Let M be a real C^2-hypersurface in an n-dimensional complex manifold X, E a holomorphic vector bundle on X, and $[D, \Omega]$ a strictly q-convex extension in M (Definition 7.1 (ii)), $1 \leq q \leq (n-1)/2$. Then*

(i) *For all r with $n - q \leq r \leq n - 1$, the restriction map*

$$(8.1) \qquad H^{n,r}_{<1/2}(\overline{\Omega}, E) \to H^{n,r}_{<1/2}(\overline{D}, E)$$

is surjective.

(ii) *If at least one of the conditions:*
   (a) $n - q + 1 \leq r \leq n - 1$
   (b) $r = n - q$ *and D is strictly q-convex in M*
   is fulfilled, then (8.1) is also injective.

(iii) *If D is strictly q-convex in M, then the restriction map*

$$(8.2) \qquad \text{germ } Z^0_{n,n-q-1}(\overline{\Omega}, E) \to \text{germ } Z^0_{n,n-q-1}(\overline{D}, E)$$

has dense image with respect to uniform convergence on $\overline{D}$.

(iv) *If D is strictly q-convex in M, then the restriction map*

$$(8.3) \qquad \text{germ } Z^{<1/2}_{n,n-q-1}(\overline{\Omega}, E) \to \text{germ } Z^{<1/2}_{n,n-q-1}(\overline{D}, E)$$

has dense image with respect to the topology of $C^{<1/2}_{n,n-q-1}(\overline{D}, E)$.

PROOF OF (i). This follows immediately from Theorem 7.10 and Lemma 6.6 (i).

PROOF OF (ii). Let U_φ, φ, c_0, c_1 be as in Definition 7.1 (ii), choose $\varepsilon > 0$ sufficiently small, set $D_- = D \setminus \{\varphi \geq c_0 - \varepsilon\}$, $\Omega_+ = \Omega \cup \{c_1 \leq \varphi < c_1 + \varepsilon\}$, and consider the sequence

$$H^{n,r}_{<1/2}(\overline{\Omega}_+, E) \xrightarrow{\alpha} \text{germ } H^{n,r}_{<1/2}(\overline{\Omega}, E) \xrightarrow{\beta} H^{n,r}_{<1/2}(\overline{\Omega}, E) \xrightarrow{\Phi}$$

$$\xrightarrow{\Phi} H^{n,r}_{<1/2}(\overline{D}, E) \xrightarrow{\gamma} \text{germ } H^{n,r}_{<1/2}(\overline{D}_-, E).$$

Then, by part (i), $\beta\alpha$ is surjective. Hence β is surjective. On the other hand, by Theorem 7.10 and Lemma 6.6 (ii), $\gamma\Phi\beta$ is injective. Together this implies that Φ is injective.

PROOF OF (iii). This follows immediately from Theorem 7.10 and Lemma 6.6 (iii).

PROOF OF (iv). Consider an open M-neighborhood U of $\overline{D}$ and a form $f \in Z^{<1/2}_{n,n-q-1}(U, E)$. Let φ, U_φ, c_0, c_1 be as in Definition 7.1 (ii). Set $D_\varepsilon = D \cup \{c_0 \le \varphi < \varepsilon\}$ and $\Omega_\varepsilon = \Omega \cup \{c_1 \le \varphi < c_1 + \varepsilon\}$. Choose $\varepsilon > 0$ so small that $D_\varepsilon \subset\subset M$, $[D_\varepsilon, \Omega_\varepsilon]$ is a strictly q-convex extension in M, and Ω_ε is strictly q-convex in M.

Then, by part (iii), the map

$$\text{germ } Z^0_{n,n-q-1}(\overline{\Omega}_\varepsilon, E) \to \text{germ } Z^0_{n,n-q-1}(\overline{D}_\varepsilon, E)$$

has dense image with respect to uniform convergence on $\overline{D}_\varepsilon$, and, by Theorem 5.2 (iv), the map

$$\text{germ } Z^{<1/2}_{n,n-q-1}(\overline{\Omega}_\varepsilon, E) \to \text{germ } Z^0_{n,n-q-1}(\overline{\Omega}_\varepsilon, E)$$

has dense image with respect to uniform convergence on $\overline{\Omega}_\varepsilon$. Therefore we can find a sequence $f'_k \in Z^{<1/2}_{n,n-q-1}(\overline{\Omega}_\varepsilon, E)$ which converges to $f|_{\overline{D}_\varepsilon}$ uniformly on $\overline{D}_\varepsilon$. Take a real C^1-function χ on M with compact support in D_ε and with $\chi = 1$ in a neighborhood of $\overline{D}$, and define by

$$f''_k = \chi f + (1 - \chi) f'_k$$

a sequence $f''_k \in C^{<1/2}_{n,n-q-1}(\overline{\Omega}_\varepsilon, E)$. Then the sequence

$$df''_k = d\chi \wedge (f - f'_k) \in Z^0_{n,n-q}(\overline{\Omega}_\varepsilon, E)$$

converges to zero uniformly on $\overline{\Omega}_\varepsilon$ and we get from Theorem 5.2 (iii) a sequence $u_k \in C^{<1/2}_{n,n-q-1}(\overline{\Omega}_\varepsilon, E)$ which converges to zero in the topology of $C^{<1/2}_{n,n-q-1}(\overline{\Omega}_\varepsilon, E)$ such that setting now $f_k = f''_k - u_k$, we obtain a sequence $f_k \in Z^{<1/2}_{n,n-q-1}(\overline{\Omega}_\varepsilon, E)$ whose restriction to $\overline{D}$

$$f_k|_{\overline{D}} = f|_{\overline{D}} - u_k|_{\overline{D}}$$

converges to $f|_{\overline{D}}$ in the topology of $C^{<1/2}_{n,n-q-1}(\overline{D}, E)$.   ∎

THEOREM 8.2. *Let M be a real C^2-hypersurface in an n-dimensional complex manifold X, E a holomorphic vector bundle on X, and $D \subseteq M$ such that $[D, M]$ is a q-convex extension in M (Definition 7.1 (i)), $1 \leq q \leq (n-1)/2$. Then:*

(i) *For all r with $n - q \leq r \leq n - 1$, the restriction map*

$$(8.4) \qquad\qquad H^{n,r}_{<1/2}(M, E) \to H^{n,r}_{<1/2}(D, E)$$

is surjective.

(ii) *If at least one of the conditions:*
   (a) $n - q + 1 \leq r \leq n - 1$

or

   (b) $r = n - q$ and D is strictly q-convex in M
is fulfilled, then the restriction map (8.4) is also injective.

PROOF. This follows by classical arguments (see, e.g., the proof of Theorem 2.8.1 in [14]) from Theorem 8.1 (i), (ii) and (iv).   ∎

THEOREM 8.3. *Let M be a real C^2-hypersurface in an n-dimensional complex manifold X, E a holomorphic vector bundle on X, D an open subset of M, ∂D the boundary of D in M, and $\overline{D} = D \cup \partial D$. Suppose ∂D is compact and, there is a $(q+1)$-convex function φ in an open M-neighborhood U of ∂D such that $D \cap U = \{\varphi < 0\}$ and $d\varphi(z) \neq 0$ for all $z \in \partial D$. Consider the sequence of restriction maps*

$$(8.5) \qquad \text{germ } H^{n,r}_{<1/2}(\overline{D}, E) \xrightarrow{\Phi_r} H^{n,r}_{<1/2}(\overline{D}, E) \xrightarrow{\psi_r} H^{n,r}_{<1/2}(\overline{D}, E).$$

Then:

(i) Φ_r *is surjective if $n - q \leq r \leq n - 1$.*

(ii) *Both Φ_r and ψ_r are isomorphisms if at least one of the conditions:*
   (a) $n - q + 1 \leq r \leq n - 1$

or

   (b) $r = n - q$ and D is strictly q-convex in M
is fulfilled.

PROOF. Choose $\varepsilon > 0$ sufficiently small, set $D_{\pm} = (D \backslash U) \cup (U \cap \{\varphi < \pm\varepsilon\})$. Then $[D_-, D]$, $[D, D_+]$, $[D_-, D_+]$ are strictly q-convex extensions in M and if D is strictly q-convex in M, then so are also D_+ and D_-. Consider the maps

$$H^{n,r}_{<1/2}(\overline{D}_+, E) \xrightarrow{\alpha_+} \text{germ } H^{n,r}_{<1/2}(\overline{D}, E)$$

and

$$H^{n,r}_{<1/2}(D, E) \xrightarrow{\alpha_-} H^{n,r}_{<1/2}(\overline{D}_-, E).$$

Then, for $n - q \leq r \leq n - 1$, by Theorem 8.1 (i), $\Phi_r \alpha_+$ is surjective and hence Φ_r is surjective. Now let at least one of the conditions (a) or (b) be fulfilled. Then, by Theorem 8.1 (ii), $\alpha_- \psi_r \Phi_r$ is injective and therefore Φ_r is injective. Moreover, then $\alpha_- \psi_r$ is an isomorphism (Theorem 8.1 (i), (ii)), α_- is an isomorphism (Theorem 8.2), and therefore ψ_r is an isomorphism.   ∎

COROLLARY 8.3 (to Theorem 5.2 (ii) and Theorem 8.3). *Let M be a real C^2-hypersurface in an n-dimensional complex manifold X, E a holomorphic vector bundle on X and D a strictly q-convex open set in M, $1 \le q \le (n-1)/2$. Then, for all r with $n - q \le r \le n - 1$,*

$$\dim \operatorname{germ} H^{n,r}_{<1/2}(\overline{D}, E) = \dim H^{n,r}_{<1/2}(\overline{D}, E) = \dim H^{n,r}_{<1/2}(D, E) < \infty.$$

If D is even completely strictly q-convex in M, then, for all r with $n - q \le r \le n - 1$,

$$\operatorname{germ} H^{n,r}_{<1/2}(\overline{D}, E) = H^{n,r}_{<1/2}(\overline{D}, E) = H^{n,r}_{<1/2}(D, E) = 0.$$

COROLLARY 8.4 (to Theorem 5.2 (ii) and Theorem 8.2). *Let M be a real C^2-hypersurface in an n-dimensional complex manifold X, E a holomorphic vector bundle on X. If M is q-convex, $1 \le q \le (n - 1)/2$, then*

$$\dim H^{n,r}_{<1/2}(M, E) < \infty \ \text{if } n - q \le r \le n - 1.$$

If M is completely q-convex, $1 \le q \le (n - 1)/2$, then

$$H^{n,r}_{<1/2}(M, E) = 0 \ \text{if } n - q \le r \le n - 1.$$

The result corresponding to Corollary 8.4 in the C^∞-case is also true. This was proved recently by Hill and Nacinovich [16], [17] using different methods (without estimates up to a boundary inside a hypersurface). Nacinovich and Hill proved this C^∞-result also for CR-manifolds of arbitrary codimension.

CHAPTER II

The q-concave case

II.0 - Introduction

In this chapter we continue our study of the Andreotti-Grauert theory in real hypersurfaces. Most of the results, we present here were announced in [24]. The first chapter was devoted to the study of the q-convex case. There we got finiteness and vanishing theorems of Andreotti-Grauert type for the $\overline{\partial}_b$-cohomology with Hölder estimates up to a boundary inside the hypersurface in high degrees. Here we are interested in the q-concave case and we provide the corresponding result on the $\overline{\partial}_b$-cohomology groups of low degree.

Analogous finiteness and vanishing theorems have been obtained recently, using different techniques, by C.D. Hill and M. Nacinovich in smooth CR manifolds of arbitrary codimension but without estimates up to a boundary inside the manifold (cf. [16], [17]).

Our method consists first in proving local homotopy formulas and then to derive, by means of a version of Grauert's bump method, global finiteness and vanishing theorems for the $\overline{\partial}_b$-cohomology with Hölder estimates. Finally we apply our results to the problem of extension of CR functions and differential forms in real hypersurfaces.

To get the local homotopy formulas we use the method initiated by A. Andreotti and C.D. Hill ([4], [5]) for the study of the tangential Cauchy-Riemann equation on hypersurfaces: by means of a jump formula we reduce the problem to a $\overline{\partial}$-problem with estimates in some wedges, which is studied in [21], [22] and [23].

In the C^∞ case local homotopy formulas are constructed by F. Trèves [30] under weaker convexity conditions on the hypersurface but whithout estimates up to a boundary.

Note that our homotopy formula – as well as the homotopy formula of Trèves [30] – holds only after shrinking the domain. In the recent work [29] of

Pervenuto alla Redazione il 14 Luglio 1994.

M.C. Shaw, for certain model domains, homotopy formulas without shrinking of the domain are constructed.

Local results on the $\overline{\partial}_b$-equation with estimates up to boundary (but without homotopy formula) are given by G.M. Henkin [10] (see also [2] and [12]) in the Hölder case and by M. Nacinovich [28] in the C^∞ case.

Our main global result (see Section 9 and 10) can be stated as follows:

THEOREM 0.1. *Let M be a real C^3 hypersurface in an n-dimensional complex manifold X, E a holomorphic vector bundle over X and $D \subset\subset \Omega \subset\subset M$ two open sets such that $D = \{\varphi < 0\}$ and $\Omega = \{\varphi < 1\}$, where φ is a real C^3 function on M which is $(q + 1)$-concave (Definition 2.4) on a neighborhood of $\Omega \backslash D$. Then the restriction map*

$$H^{n,r}_{<1/2}(\overline{\Omega}, E) \to H^{n,r}_{<1/2}(\overline{D}, E)$$

(see Section 1.1 for the definition of these cohomology groups) is an isomorphism if $0 \le r \le q - 2$ and is injective if $r = q - 1$.

This result yields an improvement of a finiteness theorem proved by C.D. Hill and M. Nacinovich (Theorem 9.10) and a vanishing theorem for the $\overline{\partial}_b$-cohomology with compact support (Corollary 9.9). From Theorem 0.1, we derive also an Hartogs-Bochner theorem for differential forms in real hypersurfaces (Theorem 11.3.1).

II.1 - Notations

Let X be an n-dimensional complex manifold and E a holomorphic vector bundle on X.

II.1.1 – Let D be an open subset with C^2-boundary in X, and let $\overline{D}$ be the closure of D in X.

We denote by $C^\alpha_{n,r}(\overline{D}, E)$ $(0 \le r \le n, \ 0 \le \alpha < 1)$ the space of continuous (if $\alpha = 0$), resp. Hölder continuous with exponent α (if $\alpha > 0$), E-valued differential forms of bidegree (n, r) on $\overline{D}$.

If $\overline{D}$ is compact, then $C^\alpha_{n,r}(\overline{D}, E)$ will be considered as *Banach space* endowed with the max-norm (if $\alpha = 0$), resp. the Hölder norm with exponent α (if $\alpha > 0$).

If $\overline{D}$ is not compact, then $C^\alpha_{n,r}(\overline{D}, E)$ will be considered as *Fréchet space* endowed with the topology defined by the Banach spaces $C^\alpha_{n,r}(\overline{W}, E)$ where W runs over all open sets $W \subseteq D$ with C^2-boundary such that the closure $\overline{W}$ of W in $\overline{D}$ is compact.

The forms which are Hölder continuous with exponent $1/2 - \varepsilon$ for all $\varepsilon > 0$ are of particular interest in this chapter. Therefore we introduce also the spaces

$$C^{<1/2}_{n,r}(\overline{D}, E) := \bigcap_{\varepsilon > 0} C^{1/2-\varepsilon}_{n,r}(\overline{D}, E).$$

These spaces will be considered as Fréchet spaces endowed with the topology defined by the topologies of the spaces $C_{n,r}^{1/2-\varepsilon}(\overline{D}, E)$, $\varepsilon > 0$, i.e. a map with values in $C_{n,r}^{1/2}(\overline{D}, E)$ is continuous if and only if it is continuous as a map with values in each $C_{n,r}^{1/2-\varepsilon}(\overline{D}, E)$, $\varepsilon > 0$.

We denote by $Z_{n,r}^{\alpha}(\overline{D}, E)$ $(0 \le r \le n$, $0 \le \alpha < 1)$ and $Z_{n,r}^{<1/2}(\overline{D}, E)$ $(0 \le \alpha < 1)$ the subspaces of closed forms in $C_{n,r}^{\alpha}(\overline{D}, E)$ resp. $C_{n,r}^{<1/2}(\overline{D}, E)$. For $1 \le r \le n$ we set

$$E_{n,r}^{<1/2}(\overline{D}, E) = Z_{n,r}^0(\overline{D}, E) \cap dC_{n,r-1}^{<1/2}(\overline{D}, E)$$

and

$$H_{<1/2}^{n,r}(\overline{D}, E) = Z_{n,r}^0(\overline{D}, E) / E_{n,r}^{<1/2}(\overline{D}, E).$$

II.1.2 – Let M be a real C^2-hypersurface in X (not necessarily closed) and D an open subset with C^2-boundary in M, and let $\overline{D}$ be the closure of D in M.

A continuous E-valued differential form on D will be called of *bidegree* (n,r), $0 \le r \le n - 1$, if it is the restriction to D of a continuous E-valued $(n-r)$-form defined in some X-neighborhood of D. Using this definition, in the same way as in Section 1.1 we define the spaces: $C_{n,r}^{\alpha}(\overline{D}, E)$, $Z_{n,r}^{\alpha}(\overline{D}, E)$ $(0 \le r \le n - 1$, $0 \le \alpha < 1)$; $C_{n,r}^{<1/2}(\overline{D}, E)$, $Z_{n,r}^{<1/2}(\overline{D}, E)$ $(0 \le r \le n - 1)$; $E_{n,r}^{<1/2}(\overline{D}, E)$, $H_{<1/2}^{n,r}(\overline{D}, E)$ $(1 \le r \le n - 1)$.

Let CR^E be the sheaf of germs of local CR sections of E over M. We denote by $H^r(M, CR^E)$ the cohomology groups of M with values in the sheaf CR^E.

II.1.3 – If we write in the definitions above a $*$ instead of r then we mean the union over all r of the corresponding spaces. For example, $C_{n,*}^{\alpha}(\overline{D}, E)$ is the union of all $C_{n,r}^{\alpha}(\overline{D}, E)$ with $0 \le r \le n$ (if D is an open subset of X) resp. $0 \le r \le n - 1$ (if $D \subseteq M$).

II.1.4 – If E is the trivial line bundle, then we omit the bundle E in these definitions above, i.e. we write $C_{n,r}^{\alpha}(\overline{D})$ instead of $C_{n,r}^{\alpha}(\overline{D}, E)$ etc.

II.1.5 – For $\xi \in X$ we denote by $T_\xi^{1,0}(X)$ the holomorphic tangent space of X at ξ, i.e. if $z_1, \ldots, z_n$ are holomorphic coordinates in a neighborhood of ξ, then $T_\xi^{1,0}(X)$ consists of all tangent vectors t of the form

$$(1.1) \qquad t = \sum_{j=1}^n t_j \frac{\partial}{\partial z_j}(\xi)$$

where $t_1, \ldots, t_n$ are complex numbers.

If M is a real C^2-submanifold of X and $\xi \in M$, then we denote by $T_\xi^{1,0}(M)$ the subspace of all vectors in $T_\xi^{1,0}(X)$ which are tangential for M.

II.1.6 – Let $\xi \in X$ and ρ a real $\mathcal{C}^2$-function defined in a neighborhood of ξ. Then we denote by $L_\xi^X(\rho)$ the Levi form of ρ at ξ, *i.e.* the hermitian form on $T_\xi^{1,0}(X)$ defined by

$$L_\xi^X(\rho)t = \sum_{j,k=1}^n \frac{\partial^2 \rho(\xi)}{\partial z_j \partial \bar{z}_k} t_j \bar{t}_k$$

if $t \in T_\xi^{1,0}(X)$ is written in the form (1.1). If M is a real $\mathcal{C}^2$-submanifold of X and $\xi \in M$, then we denote by $L_\xi^M(\rho)$ the restriction of $L_\xi^X(\rho)$ to $T_\xi^{1,0}(M)$.

II.1.7 – Let D be a relatively compact domain in X. We shall say that D is strictly q-concave in X if there exist a neighborhood $U_{\partial D}$ of ∂D and a real $\mathcal{C}^2$-function ρ defined on $U_{\partial D}$ such that $D = \{z \in U_{\partial D} \mid \rho(z) < 0\}$, $d\rho|_{\partial D} = 0$ and for each $\xi \in \partial D$, $L_\xi^{\partial D}(\rho)$ admits at least q negative eigenvalues.

II.1.8 – A collection $(U, \varphi_1, \ldots, \varphi_N)$ will be called a q-configuration in $\mathbb{C}^n$ if $U \subset \mathbb{C}^n$ is a convex domain, and $\varphi_1, \ldots, \varphi_N$ are real $\mathcal{C}^2$ functions on U satisfying the following conditions

(i) $\{z \in U \mid \varphi_1(z) = \cdots = \varphi_N(z) = 0\} \neq \emptyset$

(ii) $d\varphi_1(z) \wedge \cdots \wedge d\varphi_N(z) \neq 0$ for all $z \in U$

(iii) If $\lambda \in \Delta_{1\ldots N}$ and $\varphi_\lambda := \lambda_1 \varphi_1 + \cdots + \lambda_N \varphi_N$, then the Levi form $L_{\varphi_\lambda}^{\mathbb{C}^n}(z)$ has at least $(q+1)$ positive eigenvalues.

II.1.9 – A collection $(\xi, U, \varphi_1, \ldots, \varphi_N)$ will be called a tangential q-configuration in $\mathbb{C}^n$, $0 \le q \le n-1$, if $U \subset \mathbb{C}^n$ is a convex domain, ξ a point in U and $\varphi_1, \ldots, \varphi_N$ are real $\mathcal{C}^2$ functions on U such that $\varphi_1(\xi) = \cdots = \varphi_N(\xi) = 0$ and $\partial \varphi_1(\xi) \wedge \cdots \wedge \partial \varphi_N(\xi) \neq 0$ and the following holds:
    If $\lambda \in \Delta_{1\ldots N}$ and $\varphi_\lambda := \lambda_1 \varphi_1 + \cdots + \lambda_N \varphi_N$, then the Levi form of φ_λ at ξ restricted to the complex tangent space to $\{z \in U \mid \varphi_1(z) = \cdots = \varphi_N(z) = 0\}$ at ξ has at least q positive eigenvalues.

II.2 - $(q+1)$-concave functions

Let us first recall some definitions and results which are given in Section I.2.

DEFINITION 2.1. Let M be a real $\mathcal{C}^2$-hypersurface in some n-dimensional complex manifold X, and let q be an integer with $0 \le q \le (n-1)/2$. M will be called q-*convex-concave at a point* $\xi \in M$ if for each real $\mathcal{C}^2$-function in some X-neighborhood U of ξ with $M \cap U = \{\rho = 0\}$ and $d\rho(\xi) \neq 0$, $L_\xi^M(\rho)$ has at least q positive and q negative eigenvalues. M will be called q-convex-concave if it is q-convex-concave at each point in M.

DEFINITION 2.2. Let M be a real $\mathcal{C}^2$-hypersurface in an n-dimensional complex manifold X, and let q be an integer with $0 \le q \le (n-1)/2$. A real C^2-function φ on M will be called $(q+1)$-*convex at a point* $\xi \in M$ if there exist an X-neighborhood U of ξ and real C^2-functions $\tilde{\varphi}$, ρ_+, ρ_- on U with the following properties:

- $\tilde{\varphi} = \varphi$ on $U \cap M$
- $M \cap U = \{\rho_+ = 0\} = \{\rho_- = 0\}$
- $d\rho_\pm(\varsigma) \ne 0$ for all $\varsigma \in U$
- $\rho_+ \rho_- < 0$ on $U \setminus M$
- for each $\lambda \in [0,1]$ and for each $\varsigma \in U$, the forms $L_\varsigma^X(\lambda \tilde{\varphi} + (1-\lambda)\rho_\pm)$ have at least $(q+1)$ positive eigenvalues.

PROPOSITION 2.3 (see Proposition I.2.3). *Let M be a real C^2-hypersurface in some n-dimensional complex manifold X, and let q be an integer with $0 \le q \le (n-1)/2$. A real C^2-function φ on M is $(q+1)$-convex at a point $\xi \in M$ if and only if M is q-convex-concave at ξ and, for any real C^2-extension ψ of φ to an X-neighborhood of ξ, $L_\xi^M(\psi)$ has at least q-positive eigenvalues.*

DEFINITION 2.4. Let M be a real C^2-hypersurface in an n-dimensional complex manifold X, and let q be an integer with $0 \le q \le (n-1)/2$. A real C^2-function φ on M will be called $(q+1)$-*concave at a point* $\xi \in M$ if $-\varphi$ is $(q+1)$-convex at ξ. A $(q+1)$-*concave function* on M is, by definition, a real C^2-function on M which is $(q+1)$-concave at all points in M.

REMARK 2.5. If M is a q-convex-concave real C^2-hypersurface in an n-dimensional complex manifold X, $0 \le q \le (n-1)/2$, and ψ is a strictly plurisubharmonic function on X, then the restriction of $-\psi$ to M is $(q+1)$-concave on M is the sense of the previous definition.

DEFINITION 2.6. Let M be a real C^2-hypersurface in some complex manifold X and let φ be a real C^2-function on M. Recall that a point $\xi \in M$ is called *critical for* φ if $d\varphi(\xi) = 0$. A point $\xi \in M$ will be called *generic for* φ if ξ is non critical for φ (and hence $\varphi^{-1}(\varphi(\xi))$ is smooth in a neighborhood of ξ) and, moreover, $T_\xi^{1,0}(\varphi^{-1}(\varphi(\xi))) \ne T_\xi^{1,0}(M)$.

DEFINITION 2.7. Let M be a real C^2-hypersurface in an n-dimensional complex manifold X, and let q be an integer with $0 \le q \le (n-1)/2$. Let φ be a real C^2-function on M and $\xi \in M$ a non-critical point for φ. The function φ will be called tangential $(q+1)$-*concave at the point* ξ if for each real C^2-function ρ in some X-neighborhood U of ξ with $M \cap U = \{\rho = 0\}$ and $d\rho(\xi) \ne 0$, each real C^2-extension ψ of φ to U and each $\lambda \in [0,1]$, $L_\xi^{\varphi^{-1}(\varphi(\xi))}(\lambda \rho + (1-\lambda)\psi)$ has at least q negative eigenvalues.

REMARK 2.8. Let φ be a tangential $(q+1)$-concave function at a point ξ in M. If ξ is a generic point for φ, then there exists an X-neighborhood U of ξ such that for each defining function ρ of M on U and each C^2-extension ψ of φ to U the collection $(\xi, U, -\rho, -\psi + \psi(\xi))$ is a tangential q-configuration.

We shall now study the relations between q-concavity and tangential q-concavity.

PROPOSITION 2.9. *Let M be a real C^2-hypersurface in an n-dimensional complex manifold X, and let q be an integer with $0 \leq q \leq (n-1)/2$. Let us denote by φ a real C^2-function on M and by ξ a non critical point for φ in M.*

(i)   *If φ is tangential $(q+1)$-concave at ξ, then φ is $(q+1)$-concave at ξ.*

(ii) *If φ is $(q+1)$-concave at ξ, then φ is tangential q-concave at ξ.*

(iii) *If ξ is a non generic point for φ, then φ is tangential $(q+1)$-concave at ξ if and only if φ is $(q+1)$-concave at ξ.*

PROOF. This result is an easy consequence from the definitions and from Proposition 2.3.   ∎

DEFINITION 2.10. Let M be a real C^2-hypersurface in some n-dimensional complex manifold X, and let q be an integer with $0 \leq q \leq (n-1)/2$. A real C^2-function φ on M will be called *super $(q+1)$-concave on M* if for all $\xi \in M$ the following condition is satisfied:

If ξ is non generic for φ, then φ is $(q+2)$-concave at ξ, and if ξ is generic for φ, then φ is tangential $(q+1)$-concave at ξ.

Using Proposition I.2.7, we get

PROPOSITION 2.11. *Let M be a real C^2-hypersurface in some n-dimensional complex manifold X and let φ be a super q-concave function on M, $0 \leq q \leq (n-1)/2$. If M is q-convex-concave, then, for each compact $K \subset\subset M$, there exists $c_K > 0$ such that the following is true:*

If $\chi: \mathbb{R} \to \mathbb{R}$ is a C^2-function with $\chi'(x) > 0$ and $\chi''(x) \geq c_K \chi'(x)$ for all $x \in \varphi(K)$, then $-(\chi \circ (\varphi))$ is $(q+1)$-concave at all points in K.

DEFINITION 2.12. Let M be a real C^2-hypersurface in some n-dimensional complex manifold X, and let q be an integer with $0 \leq q \leq (n-1)/2$. A real C^2-function φ on M will be called *tangential $(q+1)$-concave on M* if the following condition is satisfied:

If ξ is non generic for φ, then φ is $(q+1)$-concave at ξ, and if ξ is generic for φ, then φ is tangential $(q+1)$-concave at ξ.

REMARK 2.13.

(i) Let φ be a tangential $(q+1)$-concave function on M and $\xi \in M$ a non critical point for φ then, for any C^2 extension ψ of φ to an X neighborhood of ξ, $L_\xi^{\varphi^{-1}(\varphi(\xi))}(\psi)$ has at least q negative eigenvalues.

(ii) One may notice that a tangential $(q+1)$-concave function is $(q+1)$-concave and that a super $(q+1)$-concave function is tangential $(q+1)$-concave.

DEFINITION 2.14. For each $\beta > 0$, let χ_β be a fixed non negative real C^∞-function on $\mathbb{R}$ such that, for all $x \in \mathbb{R}$, $\chi_\beta(x) = \chi_\beta(-x)$, $|x| \leq \chi_\beta(x) \leq |x| + \beta$,

$|\chi'_\beta(x)| \leq 1$, $\chi''_\beta(x) \geq 0$ and $\chi_\beta(x) = |x|$ if $|x| \geq \dfrac{\beta}{2}$. We set

$$\min_\beta(t, s) = \frac{t+s}{2} - \chi_\beta\left(\frac{t-s}{2}\right) \quad \text{for } t, \ s \in \mathbb{R}.$$

LEMMA 2.15. *Let* φ, ψ *be two real* $\mathcal{C}^2$-*functions defined on some complex manifold* X. *Then, for all* $\beta > 0$ *and* $z \in X$, *the following assertions hold:*

(i) $\min(\varphi(z), \psi(z)) - \beta \leq \min\limits_\beta(\varphi(z), \Psi(z)) \leq \min(\varphi(z), \psi(z))$.

(ii) $\min\limits_\beta(\varphi(z), \psi(z)) = \min(\varphi(z), \psi(z))$ *if* $|\varphi(z) - \psi(z)| \geq \beta$.

(iii) *There is a number* λ_z *with* $0 \leq \lambda_z \leq 1$, *namely* $\lambda_z := \dfrac{1}{2} - \dfrac{\chi'_\beta(\varphi(z) - \psi(z))}{2}$
such that

$$d \min_\beta(\varphi(z), \psi(z)) = \lambda_z d\varphi(z) + (1 - \lambda_z) d\psi(z).$$

(iv) *Let* $\xi \in X$, *such that* $\varphi(\xi) = \psi(\xi) = 0$, $d\varphi(\xi) \wedge d\psi(\xi) \neq 0$, *and suppose that, for all* $\lambda \in [0,1]$, $L_\xi^{\{\varphi=\psi=0\}}(\lambda\varphi + (1-\lambda)\psi)$ *has a least* q *positive eigenvalues. Then there exists an* X-*neighborhood* $U(\xi)$ *of* ξ *independent of* β *such that, for all* $z \in U(\xi)$,

$$L_z^{\{\min\limits_\beta(\varphi,\psi)=\min\limits_\beta(\varphi(z),\psi(z))\}}(\min_\beta(\varphi, \psi))$$

has at least q *positive eigenvalues.*

PROOF. We omit the proof of (i), (ii) and (iii), which is very simple. Let us prove (iv).

Choose (cf. proof of Lemma 2.4 in [21]) a continuous family T_λ, $0 \leq \lambda \leq 1$, of q-dimensional subspaces of $T_\xi^{1,0}\{\varphi = \psi = 0\}$ such that $L_z^X(\lambda\varphi + (1-\lambda)\psi)$ is positive definite on T_λ for $z = \xi$ and for all $\lambda \in [0,1]$.

Let $W(\xi)$ be an X-neighborhood of ξ so small that the preceeding property is true for all $z \in W(\xi)$.

Choose $U(\xi) \subset W(\xi)$ an X-neighborhood of ξ so small that for each $\lambda \in [0,1]$.

(a) $\tilde{T}_\lambda := T_\lambda \cap T_z^{1,0}\{\varphi = \psi = \varphi(z)\}$ has dimension q.

(b) $L_z^X(\lambda\varphi + (1-\lambda)\psi)$ is positive definite on $\tilde{T}_\lambda$.

Now, let $z \in U(\xi)$ be given. We have to find a q-dimensional subspace T_z of $T_z^{1,0}\{\min\limits_\beta(\varphi, \psi) = \min\limits_\beta(\varphi(z), \psi(z))\}$ such that $L_z^X(\min\limits_\beta(\varphi, \psi))$ is positive definite on T_z.

Let $\lambda_z \in [0,1]$ as in (iii) and set:

$$T_z := \tilde{T}_{\lambda_z}.$$

Now let $t \in \tilde{T}_{\lambda_z}$ be given, $t \neq 0$. We want to prove that

$$(2.1) \qquad L_z^X(\min_\beta(\varphi, \psi))t > 0.$$

Without lost of generality, we may assume that one can find holomorphic coordinates on $U(\xi)$. Then

$$(2.2) \qquad \begin{aligned} \frac{\partial^2 \min\limits_\beta(\varphi, \psi)}{\partial z_j \partial \overline{z}_k}(z) &= \frac{\partial}{\partial z_j}\left[\lambda_z \frac{\partial \varphi}{\partial \overline{z}_k} + (1 - \lambda_z)\frac{\partial \psi}{\partial \overline{z}_k}\right](z) \\ &= -\frac{1}{2}\chi_\beta''(\varphi(z) - \psi(z))\left(\frac{\partial \varphi}{\partial z_j}(z) - \frac{\partial \psi}{\partial z_j}(z)\right)\frac{\partial \varphi}{\partial \overline{z}_k}(z) \\ &\quad + \frac{1}{2}\chi_\beta''(\varphi(z) - \psi(z))\left(\frac{\partial \varphi}{\partial z_j}(z) - \frac{\partial \psi}{\partial z_j}(z)\right)\frac{\partial \psi}{\partial \overline{z}_k}(z) \\ &\quad + \lambda_z \frac{\partial^2 \varphi}{\partial z_j \partial \overline{z}_k}(z) + (1 - \lambda_z)\frac{\partial^2 \psi}{\partial z_j \partial \overline{z}_k}(z). \end{aligned}$$

by (iii).

Since $t \in \tilde{T}_{\lambda_z} \subset T_z^{1,0}\{\varphi = \psi = \varphi(z)\}$, we have

$$(2.3) \qquad \sum_{j=1}^n \frac{\partial \varphi}{\partial z_j}(z)t_j = \sum_{j=1}^n \frac{\partial \psi}{\partial z_j}(z)t_j = 0$$

so we deduce from (2.2) and (2.3) that

$$\begin{aligned} L_z^X(\min_\beta(\varphi, \psi))t &= \sum_{j,k=1}^n \frac{\partial^2 \min\limits_\beta(\varphi, \psi)}{\partial z_j \partial \overline{z}_k}(z)t_j \overline{t}_k \\ &= L_z^X(\lambda_z \varphi + (1 - \lambda_z)\psi)t. \end{aligned}$$

By (b), since $t \in \tilde{T}_{\lambda_z}$ and $t \neq 0$, this proves (2.1). $\qquad\blacksquare$

II.3 - Local jumps

In this section we give a local jump theorem in the q-concave situation analogous to Theorem I.3.5 which is valid in linearly q-convex open sets.

We work in $\mathbb{C}^n$ and denote by $B(z, \varsigma)$ the Bochner-Martinelli-Koppelman kernel.

If Ω is a compact oriented C^2-hypersurface with C^2 boundary $\partial\Omega$ in $\mathbb{C}^n$, then, for each continuous differential form f on Ω, resp. $\partial\Omega$, we use the abbreviations

$$(3.1) \qquad B_\Omega f(z) := \int\limits_{\varsigma \in \Omega} f(\varsigma) \wedge B(z, \varsigma), \quad z \in \mathbb{C}^n \backslash \Omega$$

and

$$(3.2) \qquad B_{\partial\Omega} f(z) := \int\limits_{\varsigma \in \partial\Omega} f(\varsigma) \wedge B(z, \varsigma), \quad z \in \mathbb{C}^n \backslash \partial\Omega.$$

We use the notation $\|f(\varsigma)\|$ to denote the norm of the differential form f at the point ς (cf. section 0.4 in [15]). Let M be a manifold, f a differential form defined on an open set D_f of M, N a submanifold of D_f and $f|_N$ the restriction of f to N, then, for $\varsigma \in N$, we shall write also $\|f(\varsigma)\|_M$ instead of $\|f(\varsigma)\|$ and $\|f(\varsigma)\|_N$ instead of $\|f|_N(\varsigma)\|$.

Let us recall the following proposition which is proved in Chapter I, Section 3.

PROPOSITION 3.1. *Let* Ω *be a compact oriented* C^2-*hypersurface with piecewise* C^2 *boundary* $\partial\Omega$ *in* $\mathbb{C}^n$. *Then*

(i) *There exists a constant* $c > 0$ *such that*

$$(3.3) \qquad \|B_\Omega f(z)\|_{\mathbb{C}^n} \leq c(1 + |\ell n \operatorname{dist}(z, \Omega)|) \max_{\varsigma \in \Omega} \|f(\varsigma)\|_\Omega$$

for all $f \in C^0_{n,*}(\Omega)$ *and* $z \in \mathbb{C}^n \backslash \Omega$. *Moreover, if* $f \in C^\alpha_{n,*}(\Omega)$ *for some* $\alpha > 0$, *then* $B_\Omega f$ *admits continuous extensions from both sides to* $\Omega \backslash \partial\Omega$.

(ii) *Let* ρ, φ *be real* C^2 *functions on* $\mathbb{C}^n$ *with* $\Omega = \{\rho = 0\} \cap \{\varphi \leq 0\}$, $d\rho(z) \neq 0$ *for* $z \in \Omega$, *and* $d\rho(z) \wedge d\varphi(z) \neq 0$ *for* $z \in \partial\Omega$. *Set*

$$(3.4) \qquad \gamma(z) = \|\partial\rho(z) \wedge \partial\varphi(z)\|_{\mathbb{C}^n}.$$

Then there exists $c > 0$ *such that*

$$(3.5) \qquad \|B_{\partial\Omega} f(z)\|_{\mathbb{C}^n} \leq c \left(\frac{\gamma(z)}{\operatorname{dist}(z, \partial\Omega)} + 1 + |\ell n \operatorname{dist}(z, \partial\Omega)| \right) \max_{\varsigma \in \partial\Omega} \|f(\varsigma)\|_\Omega$$

for all $f \in C^0_{n,*}(\Omega)$ *and* $z \in \mathbb{C}^n \backslash \partial\Omega$.

(iii) *If* $f \in C^0_{n,*}(\Omega)$ *and* $df \in C^0_{n,*}(\Omega)$, *then*

$$(3.6) \qquad dB_\Omega f + B_\Omega df = B_{\partial\Omega} f \quad on \quad \mathbb{C}^n \backslash \Omega$$

and

$$(3.7) \qquad dB_{\partial\Omega} f - B_{\partial\Omega} df = 0 \quad on \quad \mathbb{C}^n \backslash \partial\Omega.$$

(iv) *Let* $f \in C^0_{n,r}(\Omega)$ *and* $df \in C^0_{n,r+1}(\Omega)$, $0 \leq r \leq n - 1$.

    Then

(3.8) $$(-1)^{n+r} \int_{\Omega} f \wedge \varphi = \int_{\mathbb{C}^n \setminus \Omega} dB_{\Omega} f \wedge \varphi + (-1)^{n+r} \int_{\mathbb{C}^n \setminus \Omega} B_{\Omega} f \wedge d\varphi$$

for all C^{∞}-forms φ with compact support in $\mathbb{C}^n \setminus \partial\Omega$. (The integrals on the right hand side of (3.8) exist in view of estimates (3.3), (3.5) and relation (3.6)). If, moreover, f is Hölder continuous on Ω and $B_{\Omega}^+ f$, resp. $B_{\Omega}^- f$ is the continuous extension of $B_{\Omega} f$ from the left, resp. from the right to $\Omega \setminus \partial\Omega$, then

(3.9) $$(-1)^{n+r} f(z) = B_{\Omega}^+ f(z)|_{\Omega} - B_{\Omega}^- f(z)|_{\Omega}, \quad z \in \Omega \setminus \partial\Omega.$$

The jump representation (3.8), resp. (3.9), is not necessarily closed if f is closed. We shall now modify this jump to get this property. For this we have to solve a $\bar{\partial}$-equation and, since we want to preserve (at least "almost") estimate (3.3), we have to do this with appropriate uniform estimates and thus we need some convexity or concavity hypothesis on Ω (the convex case, is studied in Chapter I, here we are interested in the concave situation).

DEFINITION 3.2. An open set $D \subset\subset \mathbb{C}^n$ will be called *linearly q-concave*, $0 \le q \le n-1$, if there exist two real C^2-functions ρ and ρ_* on a neighborhood U of $\overline{D}$ such that

(i) $D = \{\rho < 0\} \cap \{\rho_* < 0\}$.

(ii) $d\rho(z) \ne 0$ if $\rho(z) = 0$, $d\rho_*(z) \ne 0$ if $\rho_*(z) = 0$ and $d\rho(z) \wedge d\rho_*(z) \ne 0$ if $\rho(z) = \rho_*(z)$.

(iii) There exist holomorphic coordinates $h_1, \ldots, h_n$ on U such that $-\rho$ is strictly convex (in the linear sense) with respect to the real coordinates $\operatorname{Re} h_1$, $\operatorname{Im} h_1, \ldots, \operatorname{Re} h_{q+1}, \operatorname{Im} h_{q+1}$.

(iv) ρ_* is convex with respect to $\operatorname{Re} h_1, \ldots, \operatorname{Im} h_n$.

REMARK 3.3. If we denote by $E = \{z \in \partial D \mid \rho(z) = 0\}$, (E, D) is the image by a holomorphic map of a local q-concave wedge in the sense of Definition 2.2 in [22]. We shall call it a *local linearly q-concave wedge*.

THEOREM 3.4. *Let (E, D) be a local linearly q-concave wedge (see Definition 3.2 and Remark 3.3), $1 \le q \le n-1$, ξ a fixed point in E and N a real C^2-hypersurface in E going through ξ, defined by $N := \{z \in E \mid \varphi(z) = 0\}$, where φ is a C^2 function defined on a neighborhood of $\overline{D}$ such that $d\varphi(z) \wedge d\rho(z) \ne 0$ for all $z \in N$.*

(i) *There exist two linear operators H and M^* from $L^1_*(D) \cap C^0_*(D)$ to $C^0_*(D)$ such that for each $f \in L^1_{n,r}(D) \cap C^0_{n,r}(D)$, $0 \le r \le q-1$, with $df \in L^1_{n,r+1}(D) \cap C^0_{n,r+1}(D)$ we have*

(3.10) $$f = dHf + Hdf + M^* f \quad \text{on } D.$$

(ii) *Moreover if f satisfies the estimate*

$$(3.11) \qquad \|f(z)\|_{\mathbb{C}^n} \leq c \left(\frac{\gamma(z)}{\mathrm{dist}(z, N)} + \frac{1}{\sqrt{\mathrm{dist}(z, N)}} \right) \ for \ z \in D$$

where $\gamma(z) = \|\partial \rho(z) \wedge \partial \varphi(z)\|_{\mathbb{C}^n}$, then the form Hf is Hölder continuous with exponent $1/2$ on $D \cup (E \backslash N)$ and

$$(3.12) \qquad \|Hf(z)\|_{\mathbb{C}^n} \leq c(1 + |\ell n \, \mathrm{dist}(z, N)|^3) \max_{\varsigma \in D} \|f(\varsigma)\|_D$$

for $z \in D$.

(iii) *There exists a real* R *and a linear operator* $\tilde{H}$ *from* $L^1_{n,*}(D) \cap C^0_{n,*}(D)$ *to* $C^0_{n,*}(D \cap B(\varsigma, R))$ *such that for each* $f \in L^1_{n,r}(D) \cap C^0_{n,r}(D)$, $0 \leq r \leq q-1$, *with* $df \in L^1_{n,r+1}(D) \cap C^0_{n,r+1}(D)$ *we have*

$$(3.13) \qquad f = d\tilde{H}f + \tilde{H}df \ on \ D \cap B(\xi, R) \ if \ 0 \leq r \leq q-2$$

$$(3.14) \qquad f = d\tilde{H}f \ on \ D \cap B(\xi, R) \ if \ r = q-1 \ and \ df = 0.$$

Moreover if f *satisfies the estimate* (3.11), *then* $\tilde{H}f$ *satisfies the estimate* (3.12) *on* $D \cap B(\xi, R)$.

PROOF. Since (E, D) is a local linearly q-concave wedge, there exists a local q-concave wedge (E', D') such that E' is defined by a strictly convex function with respect to the real coordinates $\mathrm{Re}\, z_1, \mathrm{Im}\, z_1, \ldots, \mathrm{Re}\, z_{q+1}, \mathrm{Im}\, z_{q+1}$ and a biholomorphic map h from a neighborhood $U_{\overline{D}}$ of $\overline{D}$ onto a neighborhood $U_{\overline{D}'}$ of $\overline{D}'$ such that $h(D) = D'$.

We define by H' and $(M^*)'$ the integral operators defined in section 4.4 of [22] associated to D' and for which we set here

$$\begin{cases} v_j(z, \varsigma) = 2 \dfrac{\partial \rho}{\partial \varsigma_j}(\varsigma) \ for \ 1 \leq j \leq q+1 \\[3mm] v_j(z, \varsigma) = 2 \dfrac{\partial \rho}{\partial \varsigma_j}(\varsigma) - A(\overline{\varsigma}_j - \overline{z}_j) \ for \ q+2 \leq j \leq n. \end{cases}$$

Now let $H = h^* \circ H' \circ (h^{-1})^*$ and $M^* = h^* \circ (M^*)' \circ (h^{-1})^*$, then (i) is a direct consequence of Theorem 4.4.3 in [22].

Since the singularity of the kernel associated to H' is the same in the convex and concave cases the estimate (3.12) follows from Theorem 2.1 in [8].

Let T be the Henkin operator for solving the $\overline{\partial}$-equation in $B(\xi, R)$; setting $\tilde{H} = H + TM^*$, then (iii) holds by Theorems 5.3 and 5.6 in [22]. ∎

THEOREM 3.5. *Let* (E, D) *be a local linearly* q-*concave wedge*, $0 \leq q \leq n-1$, M *a closed* C^2-*hypersurface in some neighborhood of* $\overline{D}$ *such that the intersection* $M \cap \partial D$ *is transversal. Set* $\Omega = M \cap \overline{D}$ *and* $\partial \Omega = M \cap \partial D$. *Then,* *for each $\xi \in E \cap \partial\Omega$, there exist a real R and a linear operator*

(3.15) $$S: C^0_{n,*}(\Omega) \to C^0_{n,*}((D\backslash\Omega) \cap B(\xi, R))$$

which has the following properties

(i) *There is a constant $c > 0$ with*

(3.16) $$\|Sf(z)\| \leq c(1 + |\ell n \operatorname{dist}(z, \Omega)|^3) \max_{\varsigma \in \Omega} \|f(\varsigma)\|$$

for all $f \in C^0_{n,*}(\Omega)$ and $z \in (D\backslash\Omega) \cap B(\xi, R)$. Moreover, if $f \in C^\alpha_{n,*}(\Omega)$ for some $\alpha > 0$, then Sf admits continuous extensions from both sides to $(\Omega\backslash\partial\Omega) \cap B(\xi, R)$.

(ii) *If $f \in C^0_{n,r}(\Omega)$ and $df \in C^0_{n,r+1}(\Omega)$, then*

(3.17) $$\begin{cases} dSf + Sdf = 0 & \text{if } 0 \leq r \leq q - 3 \\ \\ dSf = 0 & \text{if } r = q - 2 \text{ and } df = 0 \end{cases}$$

on $(D\backslash\Omega) \cap B(\xi, R)$.

(iii) *If $f \in C^0_{n,r}(\Omega)$ and $df \in C^0_{n,r+1}(\Omega)$ where $0 \leq r \leq q - 3$ or $f \in C^0_{n,r}(\Omega)$ and $df = 0$ if $r = q - 2$, then*

(3.18) $$(-1)^{n+r} \int_\Omega f \wedge \varphi = \int_{D\backslash\Omega} dSf \wedge \varphi + (-1)^{n+r} \int_{D\backslash\Omega} Sf \wedge d\varphi$$

for all C^∞-forms φ with compact support in $D \cap B(\xi, R)$. (The integrals on the right hand side of (3.18) exist by estimate (3.16) and relation (3.17). If moreover f is Hölder continuous on Ω and S^+f, resp. S^-f, is the continuous extension of Sf from the left, resp. the right to $(\Omega\backslash\partial\Omega) \cap B(\xi, R)$, then

(3.19) $$(-1)^{n+r} f(z) = S^+f(z)|_\Omega - S^-f(z)|_\Omega, \quad z \in (\Omega\backslash\partial\Omega) \cap B(\xi, R).$$

PROOF. We set $S = B_\Omega - \tilde{H}B_{\partial\Omega}$, where B_Ω and $B_{\partial\Omega}$ are respectively the operators defined by (3.1) and (3.2) and $\tilde{H}$ the operator in Theorem 3.4.

Then the properties of S are direct consequences from Proposition 3.1 and Theorem 3.4 (see the proof of Theorem I.3.5).   ∎

REMARK 3.6. Note that Theorem 3.5 still holds if f and df are only supposed to be bounded and piecewise continuous in $\overline{\Omega}$.

II.4 - Local homotopy formulas with Hölder estimates for the tangential Cauchy-Riemann equation

In this section we give local homotopy formulas for differential forms in q-convex-concave hypersurfaces. In Chapter I such result was given for differential forms of high degree using q-convexity properties, here the local homotopy formula is available for differential forms of small degree and related with q-concavity.

THEOREM 4.1. *Let M be a q-convex-concave real C^3-hypersurface in $\mathbb{C}^n$, $1 \le q \le \dfrac{n-1}{2}$, φ a real C^3-function on M, and $\xi \in M$ a point such that $d\varphi(\xi) = 0$ and the Hessian matrix of φ at ξ is positive definite. Set $\Omega_\tau = \{z \in M \mid \varphi(z) < \varphi(\xi) + \tau\}$ for $\tau > 0$. Then there exists $\delta_0 > 0$ such that, for all δ with $0 < \delta \le \delta_0$, there exist a real δ', $0 < \delta' < \delta$ and a continuous linear operator*

$$T: C^0_{n,r}(\overline{\Omega}_\delta) \to C^{<1/2}_{n,r-1}(\overline{\Omega}_{\delta'}) \quad 1 \le r \le q$$

such that for all $f \in C^0_{n,r}(\overline{\Omega}_\delta)$ with $df \in C^0_{n,r+1}(\overline{\Omega}_\delta)$

$$(4.1) \quad \begin{cases} \text{(i)} & f|_{\Omega_{\delta'}} = dTf - Tdf & \text{if } 1 \le r \le q-2 \\ \text{(ii)} & f|_{\Omega_{\delta'}} = dTf & \text{if } df = 0 \text{ and } r = q-1. \end{cases}$$

PROOF. Since M is q-convex-concave, we can find a $\mathbb{C}^n$-neighborhood U of ξ and real C^2-functions ρ_+, ρ_- on U such that $M \cap U = \{\rho_\pm = 0\}$, $d\rho_\pm \neq 0$ for all $z \in U$, $\rho_+ \rho_- < 0$ on $U \backslash M$, and the forms $L^{\mathbb{C}^n}_z(\rho_\pm)$, $z \in U$, have at least $q+1$ positive eigenvalues. Moreover since $d\varphi(\xi) = 0$ and the Hessian matrix of φ is positive definite, after shrinking U, we can find a real C^2-extension ψ of φ to U which is strictly convex (with respect to the real linear structure of $\mathbb{C}^n$) and such that $d\psi(\xi) = 0$. Set $D_\tau = \{z \in U \mid \psi(z) < \psi(\xi) + \tau\}$ and $D^\pm_\tau = \{z \in D_\tau \mid \rho_\pm < 0\}$ for $\tau > 0$. Then it is clear that, there exists $\delta_0 > 0$ such that if $0 < \delta \le \delta_0$, the set D_δ is linearly $(n-1)$-convex and $(\Omega_\delta, D^+_\delta)$ and $(\Omega_\delta, D^-_\delta)$ are local q-concave wedges.

For $\gamma > 0$ we denote by $B^\gamma_{n,r}(D^\pm_\tau)$, $\tau > 0$, $0 \le r \le n$, the Banach spaces of forms $f \in C^0_{n,r}(D^\pm_\tau)$ with

$$\sup_{z \in D^\pm_\tau} \|f(z)\| \, [\mathrm{dist}(z, M)]^\gamma < \infty.$$

Let $S: C^0_{n,*}(\overline{\Omega}_\delta) \to C^0_{n,*}(\overline{\Omega}_\delta \backslash \Omega_\delta)$ be the jump operator from Theorem I.3.5. Then S defines linear operators

$$S_\pm: C^0_{n,r}(\overline{\Omega}_\delta) \to \cap_{\varepsilon > 0} B^\varepsilon_{n,r}(D^\pm_\delta), \quad 0 \le r \le n-1$$

which are, for each $\varepsilon > 0$, bounded as operators with values in $B^\varepsilon_{n,r}(D^\pm_\delta)$.

By Theorem 6.7 in [22] there exist a real δ', $0 < \delta' < \delta$ and continuous operators $\tilde{H}_\pm$ between the Banach spaces $B_{n,r}^\gamma(D_\delta^\pm)$ and $C_{n,r-1}^{1/2-\gamma-\varepsilon}(\overline{D}_{\delta'}^\pm)$ for $0 \le \gamma < \frac{1}{2}$, $1 \le r \le q - 1$. Therefore, setting

$$(4.2) \qquad (-1)^{n+r-1} Tf = (\tilde{H}_+ S_+ f)|_{\overline{\Omega}_{\delta'}} - (\tilde{H}_- S_- f)|_{\overline{\Omega}_{\delta'}}$$

for $f \in C_{n,r}^0(\overline{\Omega}_\delta)$, $1 \le r \le q - 1$, we obtain operators

$$T: C_{n,r}^0(\overline{\Omega}_\delta) \to \cap_{\varepsilon>0} C_{n,r-1}^{1/2-\varepsilon}(\overline{\Omega}_{\delta'}), \qquad 1 \le r \le q - 1.$$

It remains to prove formulas (4.1). Since $\Omega_{\delta'} = M \cap D_{\delta'}$, it is sufficient to fix an orientation on $\Omega_{\delta'}$ and to prove first that, if $1 \le r \le q - 2$,

$$(4.3) \qquad \int_{\Omega_{\delta'}} f \wedge \varphi = (-1)^{n+r-1} \int_{\Omega_{\delta'}} Tf \wedge d\varphi - \int_{\Omega_{\delta'}} Tdf \wedge \varphi$$

for all C^∞-forms φ with compact support in $D_{\delta'}$. To get this, we have only to repeat the end of the proof of Theorem I.4.1.

Now let us suppose that $df = 0$ and $r = q - 1$. By formula (3.14) in Chapter I, $dS_\pm f = 0$ on $\Omega_\delta^\pm$ and by Theorem 6.7 (iv) in [22], we get $d\tilde{H}_\pm S_\pm f = S_\pm f$ on $\Omega_\delta^\pm$. Hence (ii) in (4.1) follows from Stokes' formula and Theorem I.3.5 (iii). ∎

REMARK 4.2.
(i)   From (4.1) it follows that $f|_{\Omega_{\delta'}} = dTf$ if $df = 0$, $1 \le r \le q - 1$. This result was announced by Henkin in [10].

(ii) Theorem 4.1 contains a Poincaré lemma for the tangential Cauchy-Riemann equation in q-convex-concave hypersurfaces in the case of continuous differential forms of degree less or equal to $q - 1$. For C^∞ differential forms, this Poincaré lemma was proved by Nacinovich in [27].

(iii) Mei Chi Shaw considers in [29] the special case when M is defined by $\rho(z) = \operatorname{Im} z_n - |z_1|^2 - \cdots - |z_k|^2 + |z_{k+1}|^2 + \cdots + |z_{k+\ell}|^2$, where $k + \ell \le n - 1$, $k \ge q$ and $\ell \ge q$. For this q-convex-concave hypersurface she constructs a neighborhood bases $\{\omega_\delta\}$ of 0 such that the homotopy formula (4.1) (i) holds without shrinking.

(iv) Note that for $r = q - 1$, we don't prove the homotopy formula but only a Poincaré lemma. It is proved in ([29], Theorem 4) that the homotopy formula does not hold in this case.

THEOREM 4.3. *Let M be a real C^3-hypersurface in $\mathbb{C}^n$, φ a $(q+1)$-concave C^3-function on M, $1 \le q \le (n-1)/2$, and $\xi \in M$ a point with $d\varphi(\xi) \ne 0$. Set*

$$\Omega_\tau = \{z \in M \mid \varphi(z) < \varphi(\xi) \text{ and } |\xi - z| < \tau\}$$

for $\tau > 0$.

Then there exists $\delta > 0$ such that for all β with $0 < \beta < \delta$, we can find a real number α, $0 < \alpha < \beta$, and a continuous linear operator

$$T: \mathcal{C}^0_{n,r}(\overline{\Omega}_\beta) \to \mathcal{C}^{<1/2}_{n,r-1}(\overline{\Omega}_\alpha) \quad 1 \leq r \leq q-2$$

such that for all $f \in \mathcal{C}^0_{n,r}(\overline{\Omega}_\beta)$ with $df \in \mathcal{C}^0_{n,r+1}(\overline{\Omega}_\beta)$

(4.4) $\begin{cases} \text{(i)} & f|_{\Omega_\alpha} = dTf - Tdf & \text{if } 1 \leq r \leq q-3 \\ \text{(ii)} & f|_{\Omega_\alpha} = dTf & \text{if } df = 0 \text{ and } r = q-2. \end{cases}$

PROOF. Set $U_\tau = \{z \in \mathbb{C}^n \,|\, |z - \xi| < \tau\}$ for $\tau > 0$. Then, by definition of $(q+1)$-concave functions, we have a number $\delta > 0$ and real $\mathcal{C}^2$-functions ψ, ρ_+, ρ_- on U_δ such that $\psi = \varphi$ on $M \cap U_\delta$, $M \cap U_\delta = \{\rho_\pm = 0\}$, $d\rho_\pm \neq 0$ for all $z \in U_\delta$, $\rho_+\rho_- < 0$ on $U_\delta \backslash M$, and the forms

$$L_z^{\mathbb{C}^n}(\lambda\psi + (1-\lambda)\rho_\pm), \quad z \in U_\delta, \quad 0 \leq \lambda \leq 1,$$

have at least $(q+1)$ negative eigenvalues. Since $d\varphi(\xi) \neq 0$, we may also assume that $d\psi(z) \neq 0$ for all $z \in U_\delta$. Set

$$W_\tau = \{z \in U_\tau \,|\, \psi(z) < \varphi(\xi)\} \text{ and } W_\tau^\pm = \{z \in W_\tau \,|\, \rho_\pm(z) < 0\}$$

for $0 < \tau \leq \delta$. From $\rho_+\rho_- < 0$ on $U_\delta \backslash M$, we get

$$W_\tau \backslash M = W_\tau^+ \cup W_\tau^- \text{ and } W_\tau^+ \cap W_\tau^- = \emptyset, \quad \tau > 0.$$

By the Narasimhan lemma (see Theorem 1.4.14 in [14]) and by Lemma 2.3 in [22], we may moreover assume that, for $0 < \tau < \delta$, W_τ is linearly q-concave and $(E_\tau, W_\tau^\pm)$ are local q-concave wedges, where $E_\tau = \{z \in U_\tau \,|\, \varphi(z) = \varphi(\xi)\}$.

Let α' be the real number and $S: \mathcal{C}^0_{n,*}(\overline{\Omega}_\beta) \to \mathcal{C}^0_{n,*}(W_{\alpha'} \backslash \Omega_\beta)$ be the jump operator from Theorem 3.5. By estimates (3.16), S defines linear operators

$$S_\pm: \mathcal{C}^0_{n,r}(\overline{\Omega}_\beta) \to \cap_{\varepsilon>0} B^\varepsilon_{n,r}(W_{\alpha'}^\pm), \quad 0 \leq r \leq q-2,$$

which are bounded, for each $\varepsilon > 0$, as operators values in $B^\varepsilon_{n,r}(W_{\alpha'}^\pm)$.

In view of Theorem I.6.7, there exist a real number α and linear operators

$$\tilde{H}_\pm: B^\gamma_{n,r}(W_{\alpha'}^\pm) \to \cap_{\varepsilon>0} \mathcal{C}^{1/2-\gamma-\varepsilon}_{n,r-1}(\overline{W}_\alpha^\pm), \quad 0 \leq \gamma < \frac{1}{2}, \quad 1 \leq r \leq n-1.$$

Consequently the operators $\tilde{H}_\pm S_\pm$ are bounded operators from $\mathcal{C}^0_{n,r}(\overline{\Omega}_\beta)$ into $\mathcal{C}^{<1/2}_{n,r-1}(\overline{W}_\alpha^\pm)$, $1 \leq r \leq q-2$.

Therefore, setting

(4.5) $$(-1)^{n+r-1}Tf = (\tilde{H}_+S_+f)|_{\overline{\Omega}_\alpha} - (\tilde{H}_-S_-f)|_{\overline{\Omega}_\alpha}$$

for $f \in C^0_{n,r}(\overline{\Omega}_\beta)$, $1 \le r \le n - 1$, we obtain operators

$$T: C^0_{n,r}(\overline{\Omega}_\beta) \to C^{<1/2}_{n,r-1}(\overline{\Omega}_\alpha), \quad 1 \le r \le q - 2.$$

It remains to prove the homotopy formula (4.4). Since $\Omega_\alpha = M \cap W_\alpha$, we fix an orientation on Ω_α and we prove that, for all C^∞-forms φ with compact support in W_α and $f \in C^0_{n,r}(\overline{\Omega}_\beta)$

(4.6)
$$\begin{cases} \displaystyle \int_{\Omega_\alpha} f \wedge \varphi = (-1)^{n+r-1} \int_{\Omega_\alpha} Tf \wedge d\varphi - \int_{\Omega_\alpha} Tdf \wedge \varphi & \text{if } 1 \le r \le q - 3 \\ \displaystyle \int_{\Omega_\alpha} f \wedge \varphi = (-1)^{n+r-1} \int_{\Omega_\alpha} Tf \wedge d\varphi & \text{if } df = 0 \text{ and } r = q - 2. \end{cases}$$

From (3.17) and Theorem 6.7 in [22], it follows that

$$\begin{cases} d\tilde{H}_\pm S_\pm f = S_\pm f - \tilde{H}_\pm dS_\pm f = S_\pm f + \tilde{H}_\pm S_\pm df & \text{if } 1 \le r \le q - 3 \\ d\tilde{H}_\pm S_\pm f = S_\pm f & \text{if } df = 0 \text{ and } r = q - 2 \end{cases}$$

and

$$\begin{cases} d\tilde{H}_\pm S_\pm df = -dS_\pm f & \text{if } 1 \le r \le q - 3 \\ d\tilde{H}_\pm S_\pm df = 0 & \text{if } df = 0 \text{ and } r = q - 2. \end{cases}$$

If the orientation of Ω_α is induced from the orientation of ∂W^+_α, this implies by Stokes'formula that, if $1 \le r \le q - 3$

$$(-1)^{n+r-1} \int_{\Omega_\alpha} Tf \wedge d\varphi = \int_{W_\alpha \setminus \Omega_\alpha} Sf \wedge \varphi + \int_{W^+_\alpha} \tilde{H}_+ S_+ df \wedge d\varphi + \int_{W^-_\alpha} \tilde{H}_- S_- df \wedge d\varphi$$

and

$$-\int_{\Omega_\alpha} Tdf \wedge \varphi = (-1)^{n+r} \int_{W_\alpha \setminus \Omega_\alpha} dSf \wedge \varphi - \int_{W^+_\alpha} \tilde{H}_+ S_+ df \wedge d\varphi - \int_{W^-_\alpha} \tilde{H}_- S_- df \wedge d\varphi$$

for all C^∞-forms φ with compact support in W_α. This implies that the right hand side of (4.6) is equal to

$$\int_{W_\alpha \setminus \Omega_\alpha} Sf \wedge d\varphi + (-1)^{n+r} \int_{W_\alpha \setminus \Omega_\alpha} dSf \wedge \varphi$$

if $1 \le r \le q - 3$. Hence for such r, (4.6) follows from Theorem 3.5 (iii).

Let us assume now that $df = 0$ and $r = q - 2$, then

$$(-1)^{n+r-1} \int_{\Omega_\alpha} Tf \wedge d\varphi = \int_{W_\alpha \backslash \Omega_\alpha} Sf \wedge d\varphi$$

$$= \int_{\Omega_\alpha} f \wedge \varphi$$

by (3.18). ∎

THEOREM 4.4. *Let M be a real C^3-hypersurface in $\mathbb{C}^n$, φ a real C^3-function on M and ξ a generic point for φ in M. Assume that φ is tangential $(q + 1)$-concave at ξ, $1 \le q \le (n - 1)/2$. Set $\Omega_\tau = \{z \in M \mid \varphi(z) < \varphi(\xi) \text{ and } |\xi - z| < \tau\}$ for $\tau > 0$.*

Then there exists $\delta > 0$ such that for all β with $0 < \beta < \delta$, we can find a real number α, $0 < \alpha < \beta$, and a continuous linear operator

$$T : C^0_{n,r}(\overline{\Omega}_\beta) \to C^{<1/2}_{n,r}(\overline{\Omega}_\alpha), \quad 1 \le r \le q - 1$$

such that for all closed forms $f \in C^0_{n,r}(\overline{\Omega}_\beta)$

$$f|_{\Omega_\alpha} = dTf.$$

PROOF. The tangential $(q + 1)$-concavity of φ at ξ implies by Lemma 3.1.1 in [2] that the function ψ used in the proof of Theorem 4.3 can be chosen $(q + 2)$-concave. Consequently, using the notations of the proof of Theorem 4.3, the associated set W_τ is linearly $(q + 1)$-concave and $(E_\tau, W_\tau^\pm)$ are tangential q-concave wedges for sufficiently small τ (cf. Remark 2.8 and Proposition and Definition 3.3 in [23]).

To get the theorem it remains to repeat the proof of Theorem 4.3 by using Theorem 5.2 in [23] at the place of Theorem 6.7 in [22]. ∎

REMARK 4.5. In the last section of [28], Nacinovich proved results of the same type on the local solvability of the tangential Cauchy-Riemann equation for forms which are C^∞ up to the boundary, but he did not provide homotopy formulas.

We shall complete this section with results about the Hans Lewy phenomenon in 1-convex-concave hypersurfaces of $\mathbb{C}^n$.

THEOREM 4.6. *Let M be a real C^3-hypersurface in $\mathbb{C}^n$, φ a real C^3-function on M and ξ a generic point for φ in M. We assume that φ is tangential 2-concave at ξ (which implies that M is 1-convex-concave at ξ). Let U be a neighborhood of ξ in M and f a CR function defined on $U \cap \{z \in M \mid \varphi(z) < \varphi(\xi)\}$. Then f admits a holomorphic extension to a $\mathbb{C}^n$ neighborhood of ξ.*

PROOF. Following the proof of Theorem 4.4, it is a consequence of the jump Theorem 3.5 and of the extension Theorem 5.3 in [23]. ∎

REMARK 4.8. This extension result is still true when ξ is non generic if we require that φ is 3-concave at ξ.

II.5 - First global consequences of the local homotopy formula

We denote by X an n-dimensional complex manifold and by M a real C^3-hypersurface on X.

DEFINITION 5.1. An open subset D of M will be called *strictly q-concave* (resp. *strictly super q-concave*) in M if D is q-convex-concave (as an hypersurface in X) and there exists a $(q+1)$-concave (resp. super $(q+1)$-concave) function φ of class C^3 defined in some open M-neighborhood U of the boundary ∂D of D such that $d\varphi(z)\neq 0$ for all $z \in \partial D$ and $D \cap U = \{z \in U: \varphi(z) < 0\}$.

EXAMPLES.

1) If M is q-convex-concave and compact, then D is strictly q-concave if and only if $M\backslash \overline{D}$ is strictly q-convex.

2) If M is a q-convex-concave hypersurface in X and B a strictly pseudo-concave domain in X such that M and ∂B have transversal intersection, then $M \cap B$ is strictly q-concave.

3) If Definition 5.1, the case $\partial D = \emptyset$ is possible so a q-convex-concave compact hypersurface is strictly q-concave.

THEOREM 5.2. *Let M be a real C^3-hypersurface in an n-dimensional complex manifold X, E a holomorphic vector bundle on X and D a strictly q-concave open subset in M, $1 \leq q \leq \dfrac{n-1}{2}$. Then, for all r with $1 \leq r \leq q-1$,*

(i) *There exist continuous linear operators $T: C^0_{n,r}(\overline{D}, E) \to C^{<1/2}_{n,r-1}(\overline{D}, E)$ and $K: C^0_{n,r}(\overline{D}, E) \to C^{<1/2}_{n,r}(\overline{D}, E)$ such that for all $f \in C^0_{n,r}(\overline{D}, E)$ with $df \in C^0_{n,r+1}(\overline{D}, E)$*

$$f + Kf = dTf - Tdf \quad \text{if } 0 \leq r \leq q-3$$
$$f + Kf = dTf \quad \text{if } df = 0 \text{ and } r = q-2$$
$$f + Kf = dTf \quad \text{if } df = 0 \text{ and } r \geq q-1 \text{ and if moreover } \partial D = \emptyset$$
$$\text{or } D \text{ is strictly super q-concave.}$$

(ii) $\dim H^{n,r}_{<1/2}(\overline{D}, E) < \infty$ *if $0 \leq r \leq q-2$ and if moreover $\partial D = \emptyset$ or D is strictly super q-concave then*

$$\dim H^{n,q-1}_{<1/2}(\overline{D}, E) < \infty.$$

PROOF. This theorem follows by standard arguments from Theorems 4.1, 4.3 and 4.4 (see e.g. the proofs of Theorem 7.3 in [22] and Lemma 2.3.1 in [14].   ∎

Let E be a holomorphic vector bundle over X. We denote by CR^E the sheaf of germs of local CR sections of E over M. By the same arguments as used by Nacinovich and Hill in [16] in the C^∞ case, Theorem 5.2 (ii) and Theorem 4.1 (Poincaré lemma) now imply the following

COROLLARY 5.3. *Let M be a real compact C^3-hypersurface in an n-dimensional complex manifold. Then, if M is q-convex-concave, $1 \le q \le \dfrac{n-1}{2}$,*

$$\dim H^r(M, CR^E) < \infty, \text{ for } 0 \le r \le q - 1.$$

Note also that Corollary 5.3 was announced in [10] by Henkin.

II.6 - q-concave extension elements

We first recall some definitions which are given in Section 6 of Chapter I.

DEFINITION 6.1. Let M be a real C^2-hypersurface in an n-dimensional complex manifold X, and q an integer with $1 \le q \le (n-1)/2$.

(i) An *affine q-convex configuration for M* is an ordered collection $[U, D; \rho, \rho_+, \rho_-]$ where $D \subset\subset U \subset\subset X$ are open sets ($D = \emptyset$ is possible) and ρ, ρ_+, ρ_- are real C^2-functions on U such that the following conditions are fulfilled:

   - M is closed in some neighborhood of $\overline{U}$;
   - U is biholomorphically equivalent to the ball in $\mathbb{C}^n$;
   - D is relatively compact in U, $D = \{\rho < 0\}$, $d\rho(z) \ne 0$ for $z \in \partial D$;
   - $M \cap U = \{\rho_+ = 0\} = \{\rho_- = 0\}$, $d\rho_\pm(z) \ne 0$ for all $z \in U$, and $\rho_+ \rho_- < 0$ on $U \backslash M$;
   - $d\rho(z) \wedge d\rho_\pm(z) \ne 0$ for all $z \in M \cap \partial D$;
   - for all $z \in U$ and $\lambda \in [0,1]$ the forms $L_z^X(\lambda \rho + (1 - \lambda)\rho_\pm)$ have at least $q + 1$ positive eigenvalues.

(ii) A *q-convex bump for M* is an ordered collection $[U, D_1, D_2; \rho_1, \rho_2, \rho_+, \rho_-]$ such that $[U, D_j; \rho_j, \rho_+, \rho_-]$, $j = 1, 2$, are affine q-convex configurations for M with $D_1 \subseteq D_2$.

(iii) A *q-convex extension element in M* is an ordered couple $[\Theta_1, \Theta_2]$ such that $\Theta_1 \subseteq \Theta_2$ are open subsets with C^2-boundaries in M satisfying the following condition: there exists a q-convex bump $[U, D_1, D_2; \rho_1, \rho_2, \rho_+, \rho_-]$ for M with

$$\Theta_2 = \Theta_1 \cup (M \cap D_2), \quad \Theta_1 \cap D_2 = M \cap D_1, \text{ and } (\overline{\Theta_1 \backslash D_2}) \cap (\overline{\Theta_2 \backslash \Theta_1}) = \emptyset.$$

Here $\overline{\Theta_j \backslash D_i}$ is the closure of $\Theta_j \backslash D_i$ in M, j, $i = 1, 2$.

DEFINITION 6.2. A *q-concave extension element at a point ξ in M* is an ordered collection $[\xi, \Theta_1, \Theta_2]$ such that $\Theta_1 \subset \Theta_2$ are open subsets with C^3-boundaries in M satisfying the following conditions

- $\xi \in \Theta_2 \setminus \Theta_1$;
- $[M \setminus \overline{\Theta}_2, M \setminus \overline{\Theta}_1]$ is a *q-convex extension element* in M;
- If $[U, D_2, D_1; \rho_2, \rho_1, \rho_+, \rho_-]$ is an associated convex bump, we require that $\overline{\Theta}_2 \setminus \Theta_1$ is contained in a neighborhood V_ξ of ξ such that

(i) Each continuous closed (n, r)-form on U is exact on V_ξ, for $1 \leq r \leq q - 1$.

(ii) Each continuous closed (n, r)-form on $U \cap \overline{\Theta}_i$ is exact on $V_\xi \cap \overline{\Theta}_i$, for $1 \leq r \leq q - 2$ and $i = 1, 2$.

(iii) Each continuous closed (n, r)-form on $V_\xi \cap \overline{\Theta}_1$ is exact in $W_\xi \cap \overline{\Theta}_1$, where W_ξ is a M-neighborhood of ξ containing $\overline{\Theta}_2 \setminus \Theta_1$ and $1 \leq r \leq q - 2$, and each CR-function on $V_\xi \cap \Theta_1$ can be extended to $W_\xi \supset (\overline{\Theta}_2 \setminus \Theta_1)$, if $q \geq 2$.

If moreover ρ_2 and ρ_1 are super q-concave in U and (ii) and (iii) are still true for $r = q - 1$, then $[\xi, \Theta_1, \Theta_2]$ will be called a *super q-concave extension element*.

LEMMA 6.3. *Let M be a real C^3-hypersurface in an n-dimensional complex manifold X, E a holomorphic vector bundle over X, $[\xi, \Theta_1, \Theta_2]$ a q-concave extension element at ξ in M, $1 \leq q \leq \dfrac{(n-1)}{2}$, and $\overline{\Theta}_1$, $\overline{\Theta}_2$ the closures of Θ_1, resp. Θ_2, in M.*

Then, for all r with $0 \leq r \leq q - 2$, the restriction map

$$(6.1) \qquad H^{n,r}_{<1/2}(\overline{\Theta}_2, E) \to H^{n,r}_{<1/2}(\overline{\Theta}_1, E)$$

is an isomorphism.

If moreover $[\xi, \Theta_1, \Theta_2]$ is a super q-concave extension element at ξ, (6.1) is still an isomorphism for $r = q - 1$.

PROOF. Let $[U, D_2, D_1; \rho_2, \rho_1, \rho_+, \rho_-]$ be a convex bump and V_ξ be the M-neighborhood of ξ associated to the q-concave extension element $[\xi, \Theta_1, \Theta_2]$. Since U is biholomorphically equivalent to the ball, E is holomorphically trivial over U. Since $\overline{(M \setminus \overline{\Theta}_2) \setminus D_1} \cap \overline{(\Theta_2 \setminus \Theta_1)} = \emptyset$ we can find open M-neighborhoods V' and V'' of $\overline{\Theta}_2 \setminus \Theta_1$ such that $V' \subset\subset V'' \subset\subset V_\xi$ and $\overline{V}'' \cap (M \setminus \overline{\Theta}_2) \setminus D_1) = \emptyset$. Choose a real C^1-function χ on M with $\chi = 1$ on V' and $\chi = 0$ on $M \setminus \overline{V}''$.

For $r = 0$, (6.1) is an easy consequence of Definition 6.2 (iii). We first prove the surjectivity of (6.1). Let $f_1 \in C^0_{n,r}(\overline{\Theta}_1, E)$ with $1 \leq r \leq q - 2$ be given. We have to find $u_1 \in C^{<1/2}_{n,r-1}(\overline{\Theta}_1, E)$ and $f_2 \in C^0_{n,r}(\overline{\Theta}_2, E)$ with $f_2 = f_1 - du_1$ on Θ_1. From Definition 6.2 (ii) we get $u \in C^{<1/2}_{n,r-1}(V_\xi \cap \overline{\Theta}_1, E)$ with $f_1 = du$ on $V_\xi \cap \Theta_1$. Set $u_1 = \chi u$ on $\overline{\Theta}_1$ and $f_2 = \begin{cases} f_1 - du_1 & \text{on } \overline{\Theta}_1 \\ 0 & \text{on } \overline{\Theta}_2 \setminus \Theta_1 \end{cases}$. They satisfy the required conditions.

Let us consider now the injectivity of (6.1). Let $f_2 \in C_{n,r}^0(\overline{\Theta}_2, E)$ with $1 \leq r \leq q-2$ be given. We assume that there exists $u_1 \in C_{n,r-1}^{<1/2}(\overline{\Theta}_1, E)$ such that $f_2 = du_1$, then we have to find $u_2 \in C_{n,r-1}^{<1/2}(\overline{\Theta}_2, E)$ such that $f_2 = du_2$. From Definition 6.2 (ii) we get $u \in C_{n,r-1}^{<1/2}(V_\xi \cap \overline{\Theta}_2, E)$ with $f_2 = du$ on $V_\xi \cap \Theta_2$.

If $r = 1$, $u_1 - u$ is a CR section for E on $V_\xi \cap \Theta_1$, which can be extended to $\Theta_1 \cup W_\xi$, by (iii) in Definition 6.2, in a CR section v. We set $u_2 = u_1$ on $\overline{\Theta}_1$ and $u_2 = u + v$ on $W_\xi \cap \overline{\Theta}_2$.

If $r \geq 2$, $u_1 - u$ is a continuous closed form on $V_\xi \cap \overline{\Theta}_1$. So by (iii) in Definition 6.2, $u_1 - u = dv$ on $W_\xi \cap \Theta_1$ for some $v \in C_{n,r-2}^{<1/2}(W_\xi \cap \overline{\Theta}_1, E)$. We set

$$u_2 = u_1 - d\chi v \quad \text{on } \overline{\Theta}_1$$

$$= u + d(1 - \chi)v \quad \text{on } W_\xi \cap \overline{\Theta}_2$$

then u_2 is a well-defined $(n, r-1)$-form on $\overline{\Theta}_2$, of class $C^{<1/2}$ and satisfying $f_2 = du_2$ on Θ_2.

If $[\xi, \Theta_1, \Theta_2]$ is a super q-concave extension element the previous proof is still valid for $r = q - 1$. ∎

II.7 - q-concave extensions

DEFINITION 7.1. Let M be a real C^3-hypersurface in an n-dimensional complex manifold X, and q an integer with $1 \leq q \leq \dfrac{(n-1)}{2}$. In the following definitions, for subsets W of M, we denote by $\overline{W}$ the closure of W in M, and by ∂W the boundary of W in M.

(i) A *q-concave (resp. super q-concave) extension in M* is an ordered couple $[D, \Omega]$ of open subsets $D \subset \Omega$ of M satisfying the following condition: ∂D is compact, every connected component of Ω has a non-empty intersection with D, and there exist an open M-neighborhood $U_{\partial D}$ of ∂D and a $(q + 1)$-concave (resp. super $(q + 1)$-concave) C^3-function φ defined on $U_\varphi := U_{\partial D} \cup (\Omega \backslash D)$ such that for some c_0, $c_\infty \in \mathbb{R} \cup \{\infty\}$, with $c_0 < c_\infty$,
   (a) $D \cap U_\varphi = \{\varphi < c_0\}$ and $d\varphi(z) \neq 0$ for $z \in \partial D$
   and
   (b) the sets $(\Omega \backslash D) \cap \{\varphi \leq c\}$, $c_0 \leq c < c_\infty$, are compact.

(ii) A *strictly q-concave (resp. strictly super q-concave) extension* in M is an ordered couple $[D, \Omega]$ of open subsets $D \subset \Omega$ of M such that $\overline{\Omega} \backslash D$ is compact, and there exists a $(q + 1)$-concave (resp. super $(q + 1)$-concave) C^3 function φ defined in some open M neighborhood U_φ of $\overline{\Omega} \backslash D$ such that, for some numbers $c_0 < c_1$, $D \cap U_\varphi = \{\varphi < c_0\}$, $\Omega \cap U_\varphi = \{\varphi < c_1\}$ and $d\varphi(z) \neq 0$ for $z \in \partial D \cup \partial \Omega$.
   If, moreover, this function φ can be chosen so that $d\varphi(z) \neq 0$ also for all $z \in \Omega \backslash \overline{D}$ then $[D, \Omega]$ will be called *non critical*.

(iii) M will be called *q-concave (resp. super q-concave)* if there exists a strictly q-concave (resp. strictly super q-concave) open set D in M such that $[D, M]$ is a q-concave (resp. super q-concave) extension in M.

REMARK 7.2. An ordered couple $[D, \Omega]$ is a strictly q-concave extension if and only if $[M \backslash \overline{\Omega}, M \backslash \overline{D}]$ is a strictly q-convex extension.

If M is $(q + 1)$-concave, then it is super q-concave and if M is $(q + 1)$-convex-concave and super q-concave then it is $(q + 1)$-concave.

Let φ be a $(q + 1)$-concave (resp. super $(q + 1)$-concave) function on M, $\xi \in M$ a non critical point for φ and U a M-neighborhood of ξ, then the size of the neighborhood V_ξ where one can solve the $\overline{\partial}_b$-equation on $U \cap \{\varphi < \varphi(\xi)\}$ does not depend on a small C^2-perturbation of φ in a neighborhood of ξ (cf. [22] and [23]) so using this remark, in the same way than in section I.7 we get the following lemma.

LEMMA 7.3. *Let M be a real C^3-hypersurface in a n-dimensional complex manifold X, and $[D, \Omega]$ a strictly q-concave (resp. strictly super q-concave) extension in M. Suppose additional that $[D, \Omega]$ is non critical. Then there exists a finite number of sets $\Theta_1, \ldots, \Theta_N$ such that $\Theta_1 = D$, $\Theta_N = \Omega$ and the couples $[\Theta_j, \Theta_{j+1}]$, $1 \leq j \leq N - 1$, are q-concave (resp. super q-concave) extension elements in M.*

LEMMA 7.4. *Let M be a real C^3-hypersurface in a n-dimensional complex manifold X, φ a $(q + 1)$-concave (resp. super $(q + 1)$-concave) function on M, and $\xi \in M$. Set $D_\mu = \{\varphi < \mu\}$ and suppose that, for some $\alpha < \beta$ the following conditions are fulfilled:*

- *The set $D_\beta \backslash \overline{D}_\alpha$ is relatively compact in M;*
- *$d\varphi(z) \neq 0$ for $z \in M \backslash \{\xi\}$;*
- *$\alpha < \varphi(\xi) < \beta$ and ξ is a non-degenerate critical point of φ.*

Then there exists an open M-neighborhood $W \subset\subset D_\beta \backslash \overline{D}_\alpha$ of ξ such that $[D_{\varphi(\xi)} \backslash \overline{W}, D_{\varphi(\xi)} \cup W]$ is a q-concave (resp. super q-concave) extension element in M and $[D_\alpha, D_{\varphi(\xi)} \backslash \overline{W}]$, $[D_{\varphi(\xi)} \cup W, D_\beta]$ are non-critical strictly q-concave (resp. strictly super q-concave) extensions in M.

PROOF. It is a direct consequence of Lemma 7.8 in Chapter I where we have to choose τ sufficiently small such that (i)-(iii) in Definition 6.2 are satisfied.   ∎

In the same way than for the convex case we deduce from Morse lemma, Lemma 7.3 and Lemma 7.4 the following important theorem.

THEOREM 7.5. *Let M be a real C^3-hypersurface in a n-dimensional complex manifold X, and $[D, \Omega]$ a strictly q-concave (resp. strictly super q-concave) extension in M. Then there exists a finite number of sets $\Theta_1, \ldots, \Theta_N$ such that $\Theta_1 = D$, $\Theta_N = \Omega$ and the couples $[\Theta_j, \Theta_{j+1}]$, $1 \leq j \leq N - 1$ are q-concave (resp. super q-concave) extension elements in M.*

DEFINITION 7.6. Let M be a real C^2-hypersurface in a n-dimensional complex manifold X, and q an integer with $1 \leq q \leq \dfrac{n - 1}{2}$.

A relatively compact open subset D of M with C^2-boundary will be called *tangential q-concave*, if it is q-convex-concave and if moreover it admits a defining function which is tangential $(q + 1)$-concave.

PROPOSITION 7.8. *Let M be a real C^2-hypersurface in an n-dimensional complex manifold, q an integer with $1 \le q \le \dfrac{n-1}{2}$ and D a tangential q-concave domain in M. For each X-neighborhood U of $\overline{D}$, we can find a domain $\tilde{D}$ in X such that*

(i) $\tilde{D} \cap M = D$ *and* $\tilde{D} \subset\subset U$.

(ii) $\tilde{D}$ *is a strictly q-concave domain in X (cf. 1.7).*

PROOF. Let us denote by ρ a defining function for M on a X-neighborhood W of M and by φ a tangential q-concave defining function for D. By definition of tangential q-concave functions and since ∂D is C^2 smooth, there exists a real C^2-extension ψ of φ to a X-neighborhood of ∂D such that

(a) $d\psi(\xi) \wedge d\rho(\xi) \neq 0$ for all $\xi \in \partial D$.

(b) For all $\lambda \in [0,1]$ and all $\xi \in \partial D$, $L_\xi^{\partial D}(\lambda\psi + (1-\lambda)\rho)$ has at least q positive eigenvalues.

Choose ε so small that, if we denote by D_ε the set $D_\varepsilon := \{z \in X \,|\, \psi(z) < 0, |\rho(z)| < \varepsilon\}$, we have

(α) $D_\varepsilon \subset\subset U \cap W$.

(β) The assumptions of the assertion (iv) in Lemma 2.15 are satisfied for all $\xi \in X$ such that $\varphi(\xi) = 0$ and $\rho(\xi) = \pm\varepsilon$.

(γ) The hypersurfaces $M_\varepsilon = \{z \in \overline{D}_\varepsilon \,|\, \rho(z) = \varepsilon\}$ and $M_{-\varepsilon} = \{z \in \overline{D}_\varepsilon \,|\, \rho(z) = -\varepsilon\}$ are both q-convex-concave.

Set $\tilde{D} = \{z \in U \,|\, \min_{\varepsilon/4}(-\psi, \rho + \varepsilon) > 0\} \cap \{z \in U \,|\, \min_{\varepsilon/4}(-\psi, -\rho + \varepsilon) > 0\}$.

We may notice that

$$\left\{z \in U \,\middle|\, |\rho(z)| < \frac{\varepsilon}{2}\right\} \cap \partial\tilde{D} = \{z \in U \,|\, \psi(z) = 0\} \cap \left\{z \in U \,\middle|\, |\rho(z)| < \frac{\varepsilon}{2}\right\}.$$

Consequently $\tilde{D} \cap M = D$ and since $\tilde{D} \subset D_\varepsilon$, we have by ($\alpha$), $\tilde{D} \subset\subset U$.

Moreover by (β) and (γ), $\tilde{D}$ is a strictly q-concave domain in X. ∎

II.8 - Hartogs extension for CR functions

If X is a Stein manifold of complex dimension $n \ge 2$ and K a compact subset of X, with $X \backslash K$ connected, it is well-known that every holomorphic function on $X \backslash K$ has a unique holomorphic extension to X. In fact it is sufficient for X to be completely 1-convex in the sense of Definition 5.1 in [15]. This extension property of holomorphic functions is called Hartogs phenomenon.

The following example given by Hill and Nacinovich in [16] proves that the Hartogs phenomenon fails to hold for 1-convex-concave hypersurfaces: set $M = \{z \in \mathbb{C}^3 : |z_1|^2 + |z_2|^2 |z_3|^2 = 1\}$ and $K = \{z \in M : z_3 = 0\}$. Then it is clear that M is 1-convex-concave and that the CR function $f(z) = \dfrac{1}{z_3}$ defined on $M \backslash K$ has no CR extension to M.

On the other hand, it is known that the Hartogs phenomenon holds if M is a 2-convex-concave hypersurface in a Stein manifold or if M is 1-convex-concave and K sufficiently small (see [13] and [19]).

Here we provide some new conditions on 1-convex-concave hypersurfaces such that the Hartogs phenomenon holds.

We denote by X an n-dimensional complex manifold and by M a C^3 hypersurface in X.

DEFINITION 8.1. Let $D \subset M$ be a domain and q an integer with $1 \leq q \leq \dfrac{(n-1)}{2}$. We say that *the boundary of D is strictly q-concave* (resp. *strictly super q-concave*) if ∂D is compact, the intersection of ∂D with any connected component of M is non-empty, and there exists a $(q+1)$-concave (resp. super $(q+1)$-concave) function $\varphi : U \to \mathbb{R}$ in some M-neighborhood U of ∂D such that $D \cap U = \{z \in U \mid \varphi(z) < 0\}$.

PROPOSITION 8.2. *Let $D \subset M$ be a domain of class C^3 whose boundary is strictly 2-concave or only strictly super 1-concave. Further let E be a holomorphic vector bundle over X. Then there exists a neighborhood U of ∂D such that the restriction map*

$$CR(D \cup U, E) \to CR(D \backslash U, E)$$

is an isomorphism.

PROOF. Suppose that ∂D is strictly super 1-concave. For each $\xi \in \partial D$, we set

$$V_\tau^\varepsilon(\varsigma) = \{z \in M : |z - \xi| < \tau, \ \varphi(z) < \varphi(z) - \varepsilon\}.$$

Then by Theorem 4.6 and Remark 4.7, if ε and τ are sufficiently small each CR function on $V_\tau^\varepsilon(\xi)$ extends holomorphically to an X-neighborhood U_ξ of ξ.

Using the compactness of ∂D we get $\varepsilon_0 > 0$ such that, setting $U = \left(\bigcup_{\xi \in \partial D} U_\xi \right) \cap \{z \in M \mid \varphi(z) > \varphi(\xi) - \varepsilon_0\}$, the restriction map

$$CR(D \cup U, E) \to CR(D \backslash U, E)$$

is an isomorphism. ∎

THEOREM 8.3. *Let M be a real C^3-hypersurface in a n-dimensional complex manifold and E a holomorphic vector bundle over X. If D is a domain in M*

such that $[D, M]$ is a super 1-concave extension in M, then the restriction map

$$CR(M, E) \to CR(D, E)$$

is an isomorphism.

PROOF. Following the proof of Theorem 15.2 in [15], our theorem is a consequence of Proposition 8.2 and Theorem 7.5. ∎

EXAMPLE. If $M \backslash \overline{D}$ is a completely strictly 2-convex domain in M then $[D, M]$ is a 2-concave extension and hence a super 1-concave extension in M. So Theorem 8.3 holds in this situation and this generalizes the case of a 2-convex-concave hypersurface of a Stein manifold.

As in [15], we get the following corollary.

COROLLARY 8.4. *If M is a super 1-concave C^3-hypersurface in a complex manifold X and E an holomorphic vector bundle over X, then*

$$\dim CR(M, E) < \infty.$$

Let us consider once more the example of Hill and Nacinovich ([16], Section 5).

Set $\hat{M} = \{z \in \mathbb{C}P^3 \mid |z_1|^2 + |z_2|^2 - |z_3|^2 = |z_0|^2\}$. We may notice that

$$\Sigma = \{z \in \hat{M} \mid z_3 = 0\} = \{z \in \hat{M} \mid |z_1|^2 + |z_2|^2 = |z_0|^2\}$$

and if we set

$$\tilde{M} = \hat{M} \backslash \{z \in \hat{M} \mid |z_1|^2 + |z_2|^2 \leq |z_0|^2\},$$

then $\tilde{M}$ is 1-concave but $\left\{ \left(\dfrac{z_0}{z_3} \right)^m, \ m \in \mathbb{N}^* \right\}$ is an infinite family of linearly independent CR functions on $\tilde{M}$.

This proves that Corollary 8.4 fails to be true if M is only 1-concave. Nevertheless, we will prove in Theorem 9.10, that the hypothesis of Corollary 8.4 can be weakened.

We end this section with an extension theorem, which improves Theorem 8.3 when D is supposed to be relatively compact in M.

DEFINITION 8.5. Let M be a real 1-convex-concave C^2-hypersurface in an n-dimensional complex manifold X, $n \geq 3$, and D a relatively compact domain in M. We shall say that M is a *tangential 1-concave extension* of D if there exist an open M-neighborhood $U_{\partial D}$ of ∂D and a tangential 2-concave C^2-function φ defined on $U_\varphi := U_{\partial D} \cup (M \backslash D)$ such that for some $c_0, c_\infty \in \mathbb{R} \cup \{\infty\}$ with $c_0 < c_\infty$.
(a) $D \cap U_\varphi = \{\varphi < c_0\}$ and $d\varphi(z) \neq 0$ for $z \in \partial D$
and
(b) the sets $(M \backslash D) \cap \{\varphi \leq c\}$, $c_0 \leq c < c_\infty$, are compact.

THEOREM 8.6. *Let M be a real 1-convex-concave C^2-hypersurface in an n-dimensional complex manifold X, $n \geq 3$, and D a relatively compact domain in M such that M is a tangential 1-concave extension of D. Then the restriction map*

$$CR(M, E) \rightarrow CR(D, E)$$

is an isomorphism.

PROOF. By Morse lemma we may assume that φ has only non degenerate critical points. Though we can find a sequence $(D_k)_{k \in \mathbb{N}}$ of domains in M such that $D_0 = D$, $D_k = \{\varphi < c_k\}$, $D_k \subset D_{k+1}$, D_k is tangential 1-concave in the sense of Definition 7.6 and $M = \bigcup_{k=0}^{\infty} D_k$.

Then it is sufficient to prove that the restriction maps $CR(D_{k+1}, E) \rightarrow CR(D_k, E)$, $k \in \mathbb{N}$, are isomorphisms.

Since D_k is 1-convex-concave, we can find an X-neighborhood W of D_k such that each CR function on D_k extends holomorphically to W. Choose $\alpha > 0$ sufficiently small such that $D'_k = \{\varphi < c_k - \alpha\}$ is still tangential 1-concave.

Let $V \subset\subset X$ an X-neighborhood of D_{k+1} and set $M_\varepsilon = \{z \in V \mid \rho = \varepsilon\}$, where ρ is a defining function for M. We denote by ε_0 the supremum of the absolute value of all ε such that M_ε is a q-convex-concave hypersurface.

We choose ε_k and ε_{k+1} such that $0 < \varepsilon_k < \varepsilon_{k+1} < \varepsilon_0$. Since D'_k and D_{k+1} are both tangential 1-concave, by Proposition 7.8 we can construct two strictly 1-concave domains $\tilde{D}'_k$ and $\tilde{D}_{k+1}$ in X satisfying:

i) $\tilde{D}'_k \cap M = D'_k$ and $\tilde{D}_{k+1} \cap M = D_{k+1}$;

ii) $\tilde{D}'_k \subset \{z \in X \mid |\rho(z)| < \varepsilon_k\} \cap W$
      $\tilde{D}'_{k+1} \subset \{z \in X \mid |\rho(z)| < \varepsilon_{k+1}\} \cap V$;

iii) $\tilde{D}'_k \subset\subset \tilde{D}_{k+1}$.

Moreover, one can easily deduce from the proof of Proposition 7.8 that $\tilde{D}'_k$ and $\tilde{D}_{k+1}$ can be constructed such that $\tilde{D}_{k+1}$ is a 1-concave extension of $\tilde{D}'_k$ in X. Consequently the restriction map

$$\mathcal{O}(\tilde{D}_{k+1}, E) \rightarrow \mathcal{O}(\tilde{D}'_k, E)$$

is an isomorphism.

Moreover, by definition of W, the restriction map

$$CR(W, E) \rightarrow \mathcal{O}(D_k, E)$$

is an isomorphism.

Using the fact that $\tilde{D}'_k \subset W$, we get the following commutative diagram:

$$
\begin{array}{ccc}
CR(D_{k+1}, E) & \longrightarrow & CR(D_k, E) \\
\uparrow & & \uparrow \sim \\
\mathcal{O}(\tilde{D}_{k+1}, E) \underset{\sim}{\longrightarrow} \mathcal{O}(\tilde{D}'_k, E) & \longleftarrow & \mathcal{O}(W, E)
\end{array}
$$

where all maps are restriction maps. This proves that the first line of the diagram is an isomorphism.   ∎

II.9 - Invariance of the $\overline{\partial}_b$-cohomology with respect to q-concave extensions

Similarly as in [15], we derive global results from Sections 6 and 7.

THEOREM 9.1. *Let M be a real C^3-hypersurface in an n-dimensional complex manifold X, E a holomorphic vector bundle over X, and $[D, \Omega]$ a strictly q-concave extension in M, $1 \leq q \leq \dfrac{n-1}{2}$. Then the restriction map*

$$(9.1) \qquad H^{n,r}_{<1/2}(\overline{\Omega}, E) \to H^{n,r}_{<1/2}(\overline{D}, E)$$

is an isomorphism if $0 \leq r \leq q - 2$.

If furthermore $[D, \Omega]$ is a strictly super q-concave extension in M, then (9.1) is still an isomorphism for $r = q - 1$.

PROOF. This follows immediately from Theorem 7.5 and Lemma 6.3.   ∎

THEOREM 9.2. *Let M be a real C^3-hypersurface in an n-dimensional complex manifold X, E a holomorphic vector bundle over X and D a domain in M such that $[D, M]$ is a q-concave extension in M. Then, for all r with $1 \leq r \leq q - 2$, the restriction map*

$$(9.2) \qquad H^{n,r}_{<1/2}(M, E) \to H^{n,r}_{<1/2}(\overline{D}, E)$$

is an isomorphism.

If moreover $[D, M]$ is a super q-concave extension then (9.2) is also an isomorphism for $r = q - 1$.

PROOF. In view of a Morse lemma (see [15], Appendix B), we can find a sequence $(D_k)_{k \in \mathbb{N}}$ of domains $D_k \subset M$ such that $D_0 = D$ and $[D_k, D_{k+1}]$ is a strictly q-concave extension in M. Then it remains to apply Theorem 9.1 to get the result for $1 \leq r \leq q - 2$.

In the case when $[D, M]$ is a super q-concave extension then the sequence $(D_k)_{k \in \mathbb{N}}$ can be chosen such that $[D_k, D_{k+1}]$ is a strictly super q-concave extension in M and by Theorem 9.1 this gives the result for $r = q - 1$.   ∎

Now by the same arguments as in the case of Corollary 5.3 one obtains:

COROLLARY 9.3. *Let M be a real q-concave hypersurface in an n dimensional complex manifold X and E a holomorphic vector bundle over X. Then*

$$\dim H^r(M, CR^E) < \infty \text{ for } 0 \leq r \leq q - 2$$

and if M is super q-concave, we have also

$$\dim H^{q-1}(M, CR^E) < \infty.$$

THEOREM 9.4. *Let M be a real C^3-hypersurface in an n-dimensional complex manifold X, E a holomorphic vector bundle over X and D a domain of class C^3 with strictly q-concave boundary. Then, for all r with $1 \le r \le q - 2$, the restriction map*

(9.3) $$H^{n,r}_{<1/2}(\overline{D}, E) \to H^{n,r}_{<1/2}(D, E)$$

is an isomorphism.

If moreover the boundary of D is strictly super q-concave, then (9.3) is still an isomorphism for $r = q - 1$.

PROOF. Since the boundary of D is strictly q-concave, we can find a domain $G \subset D$ such that $[G, D]$ is a strictly q-concave extension in M.

The injectivity of (9.3) follows from Theorem 9.1.

Let us consider now the surjectivity. Let $f \in Z^0_{n,r}(D, E)$ be given, then $f \in Z^0_{n,r}(\overline{G}, E)$. Though, it turns out by Theorem 9.1 that there exists $F \in Z^0_{n,r}(\overline{D}, E)$ with $f - F \in E^{<1/2}_{n,r}(\overline{G}, E)$. By Theorem 9.2 it follows that $f - F \in E^{<1/2}_{n,r}(D, E)$. ∎

THEOREM 9.5. *Let M be a real C^3-hypersurface in an n-dimensional complex manifold X, E a holomorphic vector bundle over X, and $D \subset M$ a domain such that $[D, M]$ is a q-concave extension in M, $2 \le q \le \dfrac{n-1}{2}$. Then the restriction map*

$$H^{n,q-1}_{<1/2}(M, E) \to H^{n,q-1}_{<1/2}(D, E)$$

is injective.

Moreover if $f \in Z^0_{n,q-1}(M, E)$ and $u \in C^{<1/2}_{n,q-2}(D, E)$ with $du = f$ on D, then for any neighborhood U of ∂D in M, there exists a form $v \in C^{<1/2}_{n,q-2}(M, E)$ such that $dv = f$ on M and $v = u$ on $D\backslash U$.

PROOF. It is a simple consequence of Theorem 7.5 and of the following lemma. ∎

LEMMA 9.6. *Let M be a real C^3-hypersurface in an n-dimensional complex manifold X, E a holomorphic vector bundle and $[\xi, \Theta_1, \Theta_2]$ a q-concave extension element at $\xi \in M$, $2 \le q \le \dfrac{n-1}{2}$. Let $f \in Z^0_{n,q-1}(M, E)$ such that $f = du_1$ on $\overline{\Theta}_1$, with $u_1 \in C^{<1/2}_{n,q-2}(\overline{\Theta}_1, E)$. Then, for any neighborhood U of $\overline{\Theta_2\backslash\Theta_1}$, there exists $u_2 \in C^{<1/2}_{0,q-2}(\overline{\Theta}_2, E)$ such that $du_2 = f$ on Θ_2 and $u_2 = u_1$ on $\overline{\Theta}_1\backslash U$.*

PROOF. First we notice that by Definition 6.2 (i) we can find $u \in C^{<1/2}_{n,q-2}(V_\xi, E)$ with $f = du$ on V_ξ.

If $q = 2$, $u_1 - u$ is a CR section for E on $V_\xi \cap \Theta_1$, which can be extended to $\Theta_1 \cup W_\xi$, by Definition 6.2 (iii), in a CR section v. We set $u_2 = u_1$ on $\overline{\Theta}_1$ and $u_2 = u + v$ on $W_\cap\overline{\Theta}_2$.

If $q \geq 3$, $u_1 - u$ is a continuous closed form on $V_\xi \cap \overline{\Theta}_1$. Though, by (iii) in Definition 6.2, $u_1 - u = dv$ on $W_\xi \cap \Theta_1$ for some $v \in C_{n,q-3}^{<1/2}(W_\xi \cap \overline{\Theta}_1, E)$. Choose $\chi \in C^\infty(M)$ with $\operatorname{supp} \chi \subset\subset U$ and $\chi = 1$ on $\overline{\Theta}_2 \setminus \Theta_1$. We set

$$u_2 = u_1 - d\chi v \text{ on } \overline{\Theta}_1$$

$$u_2 = u + d(1 - \chi)v \text{ on } W_\xi \cap \overline{\Theta}_2,$$

then u_2 is a well defined $(n, q - 2)$-form on $\overline{\Theta}_2$ of class $C^{<1/2}$ which satisfies $f_2 = du_2$ on Θ_2 and $u_2 = u_1$ on $\Theta_1 \setminus U$. ∎

REMARK 9.7.
(i) If $q = 1$, Theorem 9.5 is still valid. Then it is simply the uniqueness theorem for CR functions in 1-convex-concave hypersurfaces.
(ii) One can notice that Theorem 9.5 holds also if f is no more continuous but only supposed to be bounded and piecewise continuous in M (see Remark 3.6).

DEFINITION 9.8. Let M be a real C^2-hypersurface in an n-dimensional complex manifold X. M is called *completely q-convex* if M admits a $(q + 1)$-convex exhausting function.

COROLLARY 9.9. *Let M be a completely q-convex real C^3-hypersurface in an n-dimensional complex manifold X and E a holomorphic vector bundle over X, then for $0 \leq r \leq q - 1$*

$$H_c^r(M, CR^E) = 0.$$

We end this section with an improvement of Corollary 9.3 and of Theorem 5.1 in [16], in the case of hypersurfaces.

THEOREM 9.10. *Let M be a real C^3-hypersurface in an n-dimensional complex manifold X and E a holomorphic vector bundle over X. We assume that there exists a tangential q-concave domain $D \subset\subset M$ such that $[D, M]$ is a q-concave extension in M. Then*

$$\dim H^{q-1}(M, CR^E) < +\infty.$$

PROOF. By Theorem 2.4 in [17] (which is true also for C^3-manifolds) there exists a neighborhood U of M such that the restriction map

$$H^{0,q-1}(U, E) \to H^{q-1}(M, CR^E)$$

is surjective.

By Proposition 7.8, there exists a strictly q-concave domain $\tilde{D}$ in X such that $\tilde{D} \subset\subset U$ and $\tilde{D} \cap M = D$.

Let us consider the following diagram

$$0$$
$$\downarrow$$

$$
\begin{array}{ccc}
H^{0,q-1}(U,E) & \longrightarrow & H^{q-1}(M,CR^E) & \longrightarrow & 0 \\
\downarrow & & \downarrow \\
H^{0,q-1}(\tilde{D},E) & \longrightarrow & H^{q-1}(D,CR^E)
\end{array}
$$

The diagram is commutative; the first horizontal line is exact by definition of U. Since $[D,M]$ is a q-concave extension in M, by Theorem 9.5, the second vertical line is also exact. Moreover, using the strict q-concavity of $\tilde{D}$ in X, we get that

$$\dim H^{0,q-1}(\tilde{D},E) < +\infty.$$

All these facts imply that

$$\dim H^{q-1}(M,CR^E) < +\infty. \qquad \blacksquare$$

II.10 - Uniqueness theorem with boundary regularity

In this section we shall prove the following result:

THEOREM 10.1. *Let M be a real C^3-hypersurface in an n-dimensional complex manifold X, E a holomorphic vector bundle over X, and $[D,\Omega]$ a strictly q-concave extension in M, $2 \leq q \leq (n-1)/2$. Then the restriction map*

$$H^{n,q-1}_{<1/2}(\overline{\Omega},E) \to H^{n,q-1}_{<1/2}(D,\dot{E})$$

is injective.

COROLLARY 10.2. *Let M be a real C^3-hypersurface in an n-dimensional complex manifold X, E a holomorphic vector bundle over X and $D \subset M$ a domain such that $[D,M]$ is a q-convex extension in M, $2 \leq q \leq (n-1)/2$. Let $f \in Z^0_{n,q-1}(M \backslash D, E)$, which vanishes outside a compact subset of M, then we can find $u \in C^{<1/2}_{n,q-2}(M \backslash D, E)$ such that u vanishes outside a compact subset of M and $du = f$ on $M \backslash \overline{D}$.*

Following the proof of Theorem 3.1 in [26], Theorem 10.1 is a consequence of Theorem 9.5 combined with a local result on the solution of the $\overline{\partial}_b$-equation with Hölder estimates up to the boundary in small rings in M.

The remainder of this section is devoted to the description and the proof of the local result.

From now on, we consider the following situation: M is a real C^3-

hypersurface in $\mathbb{C}^n$, φ a $(q+1)$-concave C^3-function on M, $1 \leq q \leq \dfrac{n-1}{2}$, and $\xi \in M$ a point with $\varphi(\xi) = 0$ and $d\varphi(\xi) \neq 0$.

II.10.1 – Some geometry

DEFINITION 10.3. A *local linearly q-concave ring* in $\mathbb{C}^n$, $1 \leq q \leq n-1$, is a triplet (U, D, ψ) such that the following holds:

(i) $U \subset\subset \mathbb{C}^n$ is a convex open set.

(ii) ψ is a real valued C^3-function in a neighborhood of U and there exist holomorphic coordinates $h_1, \ldots, h_n$ on U such that ψ is strictly convex (in the linear sense) with respect to the real coordinates $\operatorname{Re} h_1$, $\operatorname{Im} h_1, \ldots, \operatorname{Re} h_{q+1}$, $\operatorname{Im} h_{q+1}$.

(iii) $\{\psi < 0\} \neq \emptyset$ and $\{\psi < 1\} \subset\subset U$.

(iv) $D = \{0 < \psi < 1\}$ and $d\psi(z) \neq 0$ for all $z \in \partial D$.

(v) ψ is strictly convex with respect to $\operatorname{Re} h_1$, $\operatorname{Im} h_1, \ldots, \operatorname{Re} h_n$, $\operatorname{Im} h_n$ on a neighborhood of $\{\psi \geq 1\}$.

REMARK. A local q-concave domain in the sense of Definition 2.1.2 in [26] is the image by a holomorphic map of a local linearly q-concave ring.

From Lemma 2.1.4 in [26], Definition 2.2 in this chapter and Lemma 2.3 in [22], we can easily deduce the following result:

LEMMA 10.4. *There exists a local linearly q-concave ring* $(U, D, \tilde{\varphi})$ *in* $\mathbb{C}^n$ *and real C^3-functions ρ_+, ρ_- on U such that*

(i) $M \cap \{\tilde{\varphi} < 0\} \subset U \cap \{z \in M \mid \varphi(z) > 0\}$.

(ii) *There exists a neighborhood $V \subset U \cap M$ of ξ with $V \cap \{z \in M \mid \tilde{\varphi}(z) < 0\} = V \cap \{z \in M \mid \varphi(z) > 0\}$.*

(iii) $M \cap U = \{\rho_+ = 0\} = \{\rho_- = 0\}$, $\rho_+ \rho_- < 0$ *on* $U \backslash M$.

(iv) $d\tilde{\varphi}(z) \wedge d\rho_{\pm}(z) \neq 0$ *for all* $z \in U \cap M$.

(v) *For each $\lambda \in [0, 1]$ and each $\varsigma \in U$, the forms $L_\varsigma^{\mathbb{C}^n}(\lambda\tilde{\varphi} + (1 - \lambda)\rho_{\pm})$ have at least $(q + 1)$ positive eigenvalues.*

(vi) *Setting $E = \{z \in M \mid \tilde{\varphi}(z) = 0\}$ and $D_{\pm} = D \cap \{\rho_{\pm} > 0\}$, then $(E, D_{\pm})$ are local q-concave wedges in the sense of Definition 2.2 in [22].*

Let $(U, D, \tilde{\varphi})$ be a local linearly q-concave ring given by Lemma 10.4. We shall consider more precisely the associated local q-concave wedges $(E, D_{\pm})$.

If we set $\rho_* = \tilde{\varphi} - 1$, $\rho_1 = -\tilde{\varphi}$ and $\rho_2 = -\rho_{\pm}$, then $(U, \rho_1, \rho_2, \rho_*)$ is a frame for $D_{\pm}$. Following the notations of [22], we define the following manifolds

$$S_{12} = \{z \in U \mid \rho_1(z) = \rho_2(z) = 0\}$$

$$S_{12*} = \{z \in U \mid \rho_1(z) = \rho_2(z) = \rho_*(z) = 0\}$$

$$\Gamma_{12} = \{z \in U \mid \rho_1(z) = \rho_2(z)\} \cap \{z \in U \mid \rho_*(z) \leq \rho_1(z) \leq 0\}.$$

We can notice that here $S_{12*} = \emptyset$ and consequently $E = S_{12}$ is a compact manifold. Moreover the local q-concave wedges $(E, D_\pm)$ have the following property:

LEMMA 10.5. *For each neighborhood V of the set $\check{D} = \{z \in U \mid \rho_1(z) > 0$, $\rho_2(z) > 0\}$ there exists a strictly q-convex domain G such that $\check{D} \subset\subset G \subset\subset V$ and $U_\alpha = \{z \in U \mid \tilde{\varphi}(z) < \alpha + 1\}$, $\alpha > 0$ sufficiently small, is a q-convex extension of G.*

PROOF. Without loss of generality, we may suppose that $\sup_{\overline{D}} |\rho_2| \leq 1/2$.

Let V be a fixed neighborhood of $\check{D}$. We can find real numbers $\varepsilon > 0$ and $\beta > 0$ such that, if $\tilde{\varphi}_\beta := \max_\beta(-\rho_1 - \varepsilon, -\rho_2 - \varepsilon)$ and $G := \{z \in U \mid \tilde{\varphi}_\beta(z) < 0\}$, then $\check{D} \subset\subset G \subset\subset V$, $\tilde{\varphi}_\beta$ is a strictly $(q+1)$-convex function on U which coincides with $\tilde{\varphi}$ in a neighborhood of $\{z \in U \mid \tilde{\varphi}(z) = 1\}$.

Consequently $U_\alpha = \{z \in U \mid \tilde{\varphi}_\beta(z) < \alpha + 1\}$ is a q-convex extension of G. ∎

II.10.2 – Solving the $\overline{\partial}$-equation on $(E, D_\pm)$

Since $(E, D_\pm)$ are local q-concave wedges, we shall use the methods introduced in [22] to solve the $\overline{\partial}$-equation in such sets.

We proved the existence of Leray maps for each local q-concave wedge (E', D') (see Section 2 in [22]) and we associated, to each Leray map, integral operators H, M and M^* on $C^0_{n,r}(D') \cap L^1_{n,r}(D')$, $0 \leq r \leq n$ (see Section 4 in [22] for a precise definition of these operators) such that the following holds:

Let $f \in B^\beta_{n,r}(D')$ be an (n, r)-form, $0 \leq r \leq n$, $0 \leq \beta < 1$ such that $df \in B^\beta_*(D')$. Then,

(10.1) $f = dHf + Hdf + Mf$ on D'

and if r satisfies $0 \leq r \leq q - N$, $N = \text{codim}_{\mathbb{R}} E$, then

(10.2) $f = dHf + Hdf + M^*f$ on D'.

Moreover H and M^* have the following properties: let ξ be a fixed point in E, then there exists a positive real number R such that

(i) for $0 \leq \beta \leq \frac{1}{2}$, $0 < \varepsilon \leq \frac{1}{2} - \beta$ and $1 \leq r \leq n$, H is a compact operator

between the Banach spaces $B^\beta_{n,r}(D')$ and $C^{\frac{1}{2} - \beta - \varepsilon}_{n,r-1}(\overline{D'} \cap \overline{B}(\xi, R))$.

(ii) for $\frac{1}{2} \leq \beta < 1$, $0 < \varepsilon \leq 1 - \beta$ and $1 \leq r \leq n$, H is a compact operator

between the Banach spaces $B^\beta_{n,r}(D')$ and $B^{\beta + \varepsilon - \frac{1}{2}}_{n,r-1}(D' \cap B(\xi, R))$.

(iii) M^* is a bounded operator from $B^\beta_{n,*}(D')$ into $C^1_{n,*}(B(\xi, R))$ for all $\beta < 1$.

From this integral representation, we deduced the following theorem (cf. Theorem 6.7 in [22]).

THEOREM 10.6. *Let (E', D') be a local q-concave wedge defined by a q-configuration and ξ a fixed point in E'. Set $N = \operatorname{codim}_{\mathbb{R}} E'$.*

Then there exist a positive real number R and an operator S satisfying conditions (i) and (ii) overthere, such that for each $f \in B_{n,r}^{\beta}(D')$, $0 \le \beta \le 1$, $1 \le r \le q - N$, with $df \in B_*^{\beta}(D')$ we have

(a) $f = Sdf + dSf$ on $B(\xi, R) \cap D'$, if $1 \le r \le q - N - 1$.

(b) $f = dSf$ on $D' \cap B(\xi, R)$ if $df = 0$ and $r = q - N$.

The local q-concave wedges $(E, D_{\pm})$ satisfy the hypothesis of Theorem 10.6 for $N = 2$, but since E is a compact submanifold we shall prove the following result:

PROPOSITION 10.7. *There exists an operator H satisfying conditions (i) and (ii) overthere and such that for each $f \in B_{n,r}^{\beta}(D_{\pm})$, $0 \le \beta < 1$, $1 \le r \le q - 1$, with $df \in B_*^0(D_{\pm})$ and whose support does not meet the set $\{s \in U \,|\, \tilde{\varphi}(z) = 0\}$, we have*

(a) $f = dHf + Hdf$ on D, if $1 \le r \le q - 2$

(b) $f = dHf$ on D, if $df = 0$ and $r = q - 1$.

PROOF. Our starting point is formula (10.1) and we shall prove that in our case $Mf = 0$ if $1 \le r \le q - 2$ and if $df = 0$ and $r = q - 1$.

If $1 \le r \le q - 2$, we have $Mf = M^*f$, by Lemma 4.4.2 in [22] and since M^* involves only integrals in a neighborhood of $\{z \in U \,|\, \rho_*(z) = \tilde{\varphi}(z) - 1 = 0\}$ and $\operatorname{supp} f \cap \{z \in U \,|\, \rho_*(z) = 0\} = \emptyset$, changing ρ_* to $k\rho_*$, with some constant $k > 0$, we get $M^*f = 0$.

It remains to study the case where $r = q - 1$. First consider the case where $f \in \mathcal{C}_{n,q-1}^0(\overline{D}_{\pm})$ and $df \in \mathcal{C}_{n,q}^0(\overline{D}_{\pm})$, then we have, for $z \in D$,

(10.3)
$$Mf(z) = \int_{(\varsigma, \lambda) \in S_{12} \times \Delta_{12}} f(\varsigma) \wedge \hat{L}_{12}^{\psi}(z, \varsigma, \lambda) + (-1)^{n+q-2} \int_{(\varsigma, \lambda) \in \Gamma_{12} \times \Delta_{12}} df(\varsigma) \wedge \hat{G}(z, \varsigma, \lambda)$$

where $\hat{L}_{12}^{\psi}$ and $\hat{G}$ are differential forms depending on a Leray map ψ for $(E, D_{\pm})$ (cf. Section 4.3 and 4.4 in [22]).

Moreover for each fixed $(z, \lambda) \in D \times \Delta_{12}$, the differential form $\hat{L}_{12}^{\psi}(z, \cdot, \lambda)$ is defined in a neighborhood V of $\check{D} = \{z \in U \,|\, \rho_1(z) > 0, \; \rho_2(z) > 0\}$ and $\bar{\partial}_{\varsigma}[\hat{L}_{12}^{\psi}(z, \cdot, \lambda)]_{\deg \bar{\varsigma} = n-q-1} = 0$ on V.

By Lemma 10.5, we can apply the approximation Theorem 12.11 in [15] and find a sequence $(g_k)_{k \in \mathbb{N}}$ of $\bar{\partial}$-closed forms in $U_{\alpha} = \{z \in U \,|\, \tilde{\varphi}(z) < \alpha + 1\}$ which converges uniformly to $[\hat{L}_{12}(z, \cdot, \lambda)]_{\deg \bar{\varsigma} = n-q-1}$ on S_{12}. Since U_{α} is pseudoconvex, each g_k is $\bar{\partial}$-exact on U_α, i.e. $g_k = \bar{\partial}h_k$, consequently, using the fact that $\partial S_{12} = S_{12*} = \emptyset$, we have

$$\int_{\varsigma \in S_{12}} f(\varsigma) \wedge [\hat{L}_{12}^\psi]_{\deg \bar{\varsigma} = n-q-1}(z, \varsigma, \lambda) = \lim_{k \to \infty} \int_{\varsigma \in S_{12}} f(\varsigma) \wedge \bar{\partial}h_k(\varsigma) = 0$$

when $df = 0$.

This implies

$$M f(z) = \int_{(\varsigma,\lambda) \in S_{12} \times \Delta_{12}} f(\varsigma) \wedge \hat{L}_{12}^\psi(z, \varsigma, \lambda)$$

$$= \int_{\lambda \in \Delta_{12}} \left(\int_{\varsigma \in S_{12}} f(\varsigma) \wedge [\hat{L}_{12}^\psi]_{\deg \bar{\varsigma} = n-q-1}(z, \varsigma, \lambda) \right) = 0.$$

Now, let $f \in B_{n,r}^\beta(D)$, $0 \le \beta < 1$, $0 \le r \le q - 1$, such that also $df \in B_*^\beta(D)$ and $\operatorname{supp} f \cap \{z \in U_{\bar{D}} \,|\, \rho_*(z) = 0\} = \emptyset$. Choose $\varepsilon > 0$ with $\beta + \varepsilon < 1$. Then as in [22], we can find a sequence of forms $f_\nu \in C_{n,r}^0(\bar{D})$ whose support does not meet $\{z \in U_{\bar{D}} \,|\, \rho_*(z) = 0\}$ such that the forms df_ν are also continuous on $\bar{D}$ and $f_\nu \to f$ and $df_\nu \to df$ as $\nu \to \infty$ in the space $B_*^{\beta+\varepsilon}(D)$, moreover if $df = 0$ then $df_\nu \to 0$ uniformly on $\bar{D}$.

By the properties of H, we have

$$H f_\nu \to H f \quad \text{and} \quad H df_\nu \to H df.$$

By (10.1), we have

$$f_\nu = dH f_\nu + H df_\nu + M f_\nu$$

and if $r \le q - 2$, $M f_\nu = 0$.

Using (10.3) and the continuous case, we get, for $r = q - 1$

$$M f_\nu(z) = \int_{(\varsigma,\lambda) \in S_{12} \times \Delta_{12}} f_\nu(\varsigma) \wedge \hat{L}_{12}^\psi(z, \varsigma, \lambda)$$

$$+ (-1)^{n+q-2} \int_{(\varsigma,\lambda) \in \Gamma_{12} \times \Delta_{12}} df_\nu(\varsigma) \wedge \hat{G}(z, \varsigma, \lambda)$$

$$= \int_{\lambda \in \Delta_{12}} \lim_{k \to \infty} \int_{\varsigma \in S_{12}} df_\nu(\varsigma) \wedge h_k(\varsigma)$$

$$+ (-1)^{n+q-2} \int_{(\varsigma,\lambda) \in \Gamma_{12} \times \Delta_{12}} df_\nu(\varsigma) \wedge \hat{G}(z, \varsigma, \lambda).$$

Hence $|M f_\nu(z)| \leq C \|df_\nu\|_0$ for each $z \in D$ and therefore $M f_\nu \to 0$ as $\nu \to \infty$, if $df = 0$.

Finally, we obtain

$$f = dHf + Hdf \quad \text{on } D, \text{ if } 0 \leq r \leq q - 2$$

$$f = dHf \quad \text{on } D, \text{ if } df = 0 \text{ and } r = q - 1. \qquad \blacksquare$$

II.10.3 – A jump theorem for $(U, D, \tilde{\varphi})$

Let $(U, D, \tilde{\varphi})$ be a local linearly q-concave ring given by Lemma 10.4. We set $\Omega = D \cap M$.

PROPOSITION 10.8. *There exists a linear operator*

$$S: \mathcal{C}^0_{n,*}(\overline{\Omega}) \to \mathcal{C}^0_{n,*}(D \backslash \Omega)$$

such that

(i) *There is a constant $c > 0$ with*

$$\|Sf(z)\| \leq c(1 + |\ell n \operatorname{dist}(z, \Omega)|^3) \max_{\varsigma \in \Omega} \|f(\varsigma)\|$$

   for all $f \in \mathcal{C}^0_{n,*}(\Omega)$ and $z \in D \backslash \Omega$.

(ii) *If $f \in \mathcal{C}^0_{n,r}(\overline{\Omega})$ and $df \in \mathcal{C}^0_{n,r+1}(\overline{\Omega})$, then*

$$\begin{cases} dSf + Sdf = 0 & \text{if } 0 \leq r \leq q - 2 \\ dSf = 0 & \text{if } r = q - 1 \text{ and } df = 0 \end{cases}$$

   on $D \backslash \Omega$.

(iii) *If $f \in \mathcal{C}^0_{n,r}(\Omega)$ and $df \in \mathcal{C}^0_{n,r+1}(\Omega)$, for $0 \leq r \leq q - 2$ or $df = 0$ if $r = q - 1$, then for all C^∞-forms φ with compact support in D*

$$(-1)^{n+r} \int_\Omega f \wedge \varphi = \int_{D \backslash \Omega} dSf \wedge \varphi + (-1)^{n+r} \int_{D \backslash \Omega} Sf \wedge d\varphi.$$

PROOF. We set $S = B_\Omega - \tilde{H} B_{\partial \Omega}$, where B_Ω and $B_{\partial \Omega}$ are respectively the operators defined by (3.1) and (3.2) and $\tilde{H}$ is the operator in Theorem 3.4. Using Theorem 2.2.4 and Proposition 2.2.8 in [26], we can notice that (3.13) is still valid for $r = q - 1$ and (3.14) for $r = q$, which ends the proof. $\blacksquare$

REMARK 10.9. Let $f \in \mathcal{C}^0_{n,r}(\overline{\Omega})$, $0 \leq r \leq q - 1$, be a closed form such that $\operatorname{supp} f$ does not meet $\{z \in U_{\overline{D}} \mid \tilde{\varphi}(z) = 1\}$, then there exists a differential form $F \in \mathcal{C}^0_{n,r}(D \backslash \Omega)$ such that

(i)   There is a constant $c > 0$ with

$$\|F(z)\| \leq c(1 + |\ell n \text{ dist}(z, \Omega)|^3) \max_{\varsigma \in \Omega} \|f(\varsigma)\|$$

for $z \in D \backslash \Omega$.

(ii) $dF = 0$ in $D \backslash \Omega$ and $\text{supp } F \cap \{z \in U_{\overline{D}} \,|\, \tilde{\varphi}(z) = 1\} = \emptyset$.

(iii) For all C^∞-forms φ with compact support in D

$$(-1)^{n+r} \int_\Omega f \wedge \varphi = \int_{D \backslash \Omega} dF \wedge \varphi + (-1)^{n+r} \int_{D \backslash \Omega} F \wedge d\varphi.$$

In fact F is given by

$$F = \chi B_\Omega f - \tilde{H}(\overline{\partial}(\chi B_\Omega f)) + \overline{\partial} g$$

where $\chi \in \mathcal{D}(\mathbb{C}^n)$ is equal to 1 in a neighborhood of $\text{supp } f$, $\text{supp } \chi \subset\subset \{z \in U_{\overline{D}} \,|\, \tilde{\varphi}(z) < 1\}$ and $g \in C^\infty_{n,r-1}(\mathbb{C}^n)$ satisfies $\overline{\partial} g = \tilde{H}(\overline{\partial}(\chi B_\Omega f))$ on $\mathbb{C}^n \backslash \{z \in U_{\overline{D}} \,|\, \tilde{\varphi}(z) \leq 1 - \alpha\}$ for $\alpha > 0$ such that $\text{supp } \chi \subset\subset \{z \in U_{\overline{D}} \,|\, \tilde{\varphi}(z) < 1 - \alpha\}$.

II.10.4 – The local result in small rings in M

We can now give the local result on M.

THEOREM 10.10. *Let M be a real C^3-hypersurface in $\mathbb{C}^n$, φ a $(q+1)$-concave C^3 function on M, $1 \leq q \leq (n-1)/2$, and $\xi \in M$ a point with $\varphi(\xi) = 0$ and $d\varphi(\xi) \neq 0$. Let $(U, D, \tilde{\varphi})$ be the local q-concave ring given in Lemma 10.4. Let $\Omega = \{z \in M \,|\, 0 < \tilde{\varphi}(z) < 1\} = D \cap M$.*

Then there exists a continuous linear operator

$$T: C^0_{n,r}(\overline{\Omega}) \to C^{<1/2}_{n,r-1}(\overline{\Omega})$$

such that for all $f \in C^0_{n,q-1}(\overline{\Omega})$ with $df = 0$ and $\text{supp } f \subset \{z \in M \,|\, \tilde{\varphi}(z) < 1\}$, we have

$$f = dTf \text{ on } \Omega.$$

PROOF. We denote by $(E, D_\pm)$ the two local q-concave wedges associated to the local q-concave ring $(U, D, \tilde{\varphi})$ and to the hypersurface M by Lemma 10.4.

Let $f \in C^0_{n,q-1}(\overline{\Omega})$ with $df = 0$ and $\text{supp } f \subset \{z \in M \,|\, \tilde{\varphi}(z) < 1\}$. By Remark 10.9 there exists $F \in C^0_{n,q-1}(D \backslash \Omega)$ such that if $F_\pm = F|_{D_\pm}$, then $F_\pm \in \bigcap_{0 < \beta < 1} B^\beta_{n,q-1}(D_\pm)$ and $dF_\pm = 0$ on $D_\pm$.

Using Proposition 10.7 and setting $g = HF_+ - HF_-$ we have $f = dg$ on Ω and $g \in C^{<1/2}_{n,r-1}(\overline{\Omega})$. ∎

II.11 - Applications

II.11.1 – Jump theorem

In this section X is an oriented, n-dimensional complex manifold, M is an oriented, closed real C^3-hypersurface in X, V is a real C^1-hypersurface in M, E is a holomorphic vector bundle over X and E^* is the dual of E.

DEFINITION 11.1.1. A form $f \in C^0_{n,r}(V, E)$, $0 \le r \le n - 2$, is *locally the jump of continuous CR forms* if, for every point $z \in V$, there exists a neighborhood U of z in M, such that $U \backslash V$ has exactly two connected components U^+ and U^-, and differential CR forms $F^\pm \in C^0_{n,r}(U^\pm \cup (U \cap V), E)$ with $f|_{V \cap U} = F^+|_{V \cap U} - F^-|_{V \cap U}$.

REMARK 11.1.2. If $f \in C^0_{n,r}(V, E)$, $0 \le r \le n - 2$, is locally the jump of continuous CR forms then f is CR on V.

DEFINITION 11.1.3. Let $U \subset M$ be an open set and Φ the family of all subsets of U which are closed with respect to M. Then, we denote by $H^{n,r}_\Phi(U, E)$ the $\overline{\partial}_b$-cohomology group of bidegree (n, r), with coefficients in E and supports in Φ.

Following the same methods as in ([25], Section 2) we get:

THEOREM 11.1.4. *Let $U \subset M$ be a neighborhood of V such that for some q, $0 \le q \le n - 2$,*

$$(11.1.1) \qquad H^{n,r+1}_\Phi(U, E) = 0, \quad 0 \le r \le q,$$

and $U \backslash V$ has exactly two connected components U^+ and U^-. Then for each $f \in C^0_{n,r}(V, E)$, $0 \le r \le q$, which is locally the jump of continuous CR forms there exists a differential form $F \in Z^0_{n,r}(M \backslash V, E)$ such that

(i) $F \equiv 0$ *in a neighborhood of $X \backslash U$;*

(ii) $F|_{U^\pm}$ *extends continuously to $U^\pm \cup (U \cap V)$ in $F^\pm$;*

(iii) $f = F^+|_V - F^-|_V$.

DEFINITION 11.1.5. If a differential form $f \in C^0_{n,r}(V, E)$, $0 \le r \le n - 2$, satisfies the conclusion of Theorem 11.1.4, we shall say that f is *a global C^0-jump of CR forms on M with support in U*.

COROLLARY 11.1.6. *Let $U \subset M$ be a neighborhood of V such that $[M \backslash U, M]$ is a $(q + 2)$-concave extension in M and $U \backslash V$ has exactly two connected components. Then each $f \in C^0_{n,r}(V, E)$, $0 \le r \le q$, which is locally the jump of continuous CR forms, is a global C^0-jump of CR forms on M with support in U.*

PROOF. By Theorem 9.5 and Remark 9.7 (ii), condition (11.1.1) is fulfilled and the statement follows from Theorem 11.1.4.   ∎

COROLLARY 11.1.7. *Suppose V is compact and for some q, $0 \le q \le \dfrac{n-5}{2}$, M is completely $(q+2)$-convex. Then there exists a neighborhood $U \subset\subset M$ of V such that each $f \in \mathcal{C}^0_{n,r}(V, E)$, $0 \le r \le q$, which is locally the jump of continuous CR forms is a global $\mathcal{C}^0$-jump of CR forms on M with support in U.*

PROOF. Since M is completely $(q+2)$-convex, it admits a $(q+3)$-convex exhausting function ρ. As V is compact, we can find $t_0 \in \mathbb{R}$ with $V \subset U = \{z \in M \mid \rho(z) < t_0\}$. By definition of ρ, M is a $(q+2)$-concave extension of $X \backslash U$ and the assertion follows from Corollary 11.1.6. ∎

II.11.2 – Extension of CR functions in real hypersurfaces

In this section we shall study the extension of CR functions defined on the boundary of a domain D in a real hypersurface M of a complex manifold X. The same problem was studied in ([25], Section 5) in the case when D is a domain in a complex manifold.

THEOREM 11.2.1. *Let E be a holomorphic vector bundle over an n-dimensional complex manifold X, $n \ge 5$, M a closed real $\mathcal{C}^3$-hypersurface in X, $D \subset M$ an open set with $\mathcal{C}^1$ boundary (not necessarily compact) and $\Gamma \subset M$ a closed set such that the following two conditions are fulfilled:*

(i) *Γ has a non-empty intersection with each connected component of $M \backslash D$.*

(ii) *$\Gamma \cap \partial D = \emptyset$ and M is a 2-concave extension of Γ. Then, for each CR section $f \in \mathcal{C}^0(\partial D, E)$, which is locally the jump of continuous CR functions, there exists $F \in \mathcal{C}^0(\overline{D}, E)$, CR in D with $F|_{\partial D} = f$.*

PROOF. Since f is locally the jump of continuous CR functions, using Corollary 11.1.6 with $U = M \backslash \Gamma$ we get a CR function $F \in \mathcal{C}^0(M \backslash \partial D, E)$ such that

(1) $F \equiv 0$ in a neighborhood of Γ.

(2) $F|_D$ extends continuously to $\overline{D}$ and $F|_{M \backslash \overline{D}}$ extends continuously to $M \backslash D$. Let us denote by F^+ and F^- the previous extensions, then

$$F^+|_{\partial D} - F^-|_{\partial D} = f.$$

Since M is a 2-concave extension of Γ, $M \backslash \Gamma$ is 2-convex-concave. So we deduce from (i), (1) and from the uniqueness of CR functions in 2-convex-concave hypersurfaces that $F|_{M \backslash \overline{D}} \equiv 0$. Consequently by (2), F^+ gives the desired extension. ∎

COROLLARY 11.2.2. *Let E be a holomorphic vector bundle over an n-dimensional complex manifold X, $n \ge 5$, M a closed real $\mathcal{C}^3$-hypersurface in X, $\Omega \subset\subset M$ an open set with $\mathcal{C}^1$ boundary and $K \subset \overline{\Omega}$ a closed set such that*

(i) $M\backslash\Omega$ *does not contain connected components which are relatively compact in* M.

(ii) M *is a 2-convex extension of* K.

   Then, for each CR section $f \in C^\alpha(\partial D\backslash K, E)$, $\alpha > 0$, *there exists* $F \in \bigcap_{\varepsilon>0} C^{(\alpha/2)-\varepsilon}(\overline{\Omega}\backslash K, E)$, *CR in* Ω *with* $F|_{\partial\Omega\backslash K} = f$.

   PROOF. It is sufficient to repeat the proof of Corollary 5.2 [25] in the setting of real hypersurfaces and to use the local jump Theorem 5.2 of [6].   ∎

   REMARK 11.2.3.

   1) It is easy to see (by uniqueness of CR functions in 2-convex-concave hypersurfaces) that Corollary 11.2.2 remains valid if we replace condition (ii) by:

   (ii′) There exists a basis $\mathcal{U}$ of neighborhoods of K in M such that for each $U \in \mathcal{U}$, M in a 2-convex extension of U.

   As it was already done in [20], let us notice that (ii′) is fulfilled if M is a 2-convex-concave hypersurface in a Stein manifold X and K the intersection of M with a Stein compactum S such that both $X\backslash M$ and $S\backslash M$ have exactly two connected components.

   2) When $K = \emptyset$, condition (ii) in Corollary 11.2.2 means that M is completely 2-convex and the Corollary 11.2.2 is a special case of the Hartogs-Bochner theorem for hypersurfaces given in ([20], Section 3) if one considers smooth functions.

II.11.3 – Hartogs-Bochner phenomenon for differential forms in real hypersurfaces

   Our purpose is the generalization to real hypersurfaces of the Hartogs-Bochner phenomenon for differential forms. This phenomenon was studied in complex manifold first by Kohn and Rossi [18].

   THEOREM 11.3.1. *Let* E *be a holomorphic vector bundle over an* n-*dimensional complex manifold* X, $n \geq 5$, M *a closed* C^3 *real hypersurface and* $D \subset\subset M$ *a relatively compact open set in* M *with* C^3 *boundary such that the following two conditions are fulfilled:*

(i) M *is completely* $(q+2)$-*convex,* $0 \leq q \leq \dfrac{n-5}{2}$;

(ii) $[D, M]$ *is a* $(q+1)$-*convex extension in* M.

   Then for each $f \in C^0_{n,r}(\partial D, E)$, $0 \leq r \leq q$, *which is locally the jump of continuous CR forms, there exist a CR form* $F \in C^0_{n,r}(\overline{D}, E)$ *and a form* $u \in \bigcap_{\varepsilon>0} C^{(1/2)-\varepsilon}_{n,r-1}(M\backslash D, E)$ *such that* $\overline{\partial}u \in C^0_{n,r}(M\backslash D, E)$ *and* $f = F|_{\partial D} - \overline{\partial}u|_{\partial D}$.

PROOF. It follows from Corollary 11.1.7 and (i) that $f \in C^0_{n,r}(\partial D, E)$ which is locally the jump of continuous CR forms is a global C^0 jump of CR forms. Therefore there exists $F \in C^0_{n,r}(M \setminus \partial D, E)$ such that

1) $F \equiv 0$ outside a compact set of M.

2) $F|_D$ extends continuously to $\overline{D}$ in F^+ and $F|_{M \setminus \overline{D}}$ extends continuously to $M \setminus D$ in F^-.

3) $f = F^+|_{\partial D} - F^-|_{\partial D}$.

Applying Corollary 10.2 to F^-, we obtain $u \in C^{<1/2}_{n,r-1}(M \setminus D, E)$ such that $\overline{\partial} u = F^-$. Then by 3) $f = F^+|_{\partial D} - \overline{\partial} u|_{\partial D}$. ∎

REMARK 11.3.2.

1) In real hypersurfaces Theorem 11.3.1 corresponds to Theorem 4.1 in [25] with $K = \emptyset$.

2) If $q = 0$, M is supposed to be completely 2-convex and consequently 2-convex-concave. Though, in this case condition (ii) is satisfied by any relatively compact C^3 domain D in M.

BIBLIOGRAPHY

[1] L. AIZENBERG - SH. DAUTOV, *Differential forms orthogonal to holomorphic functions or forms, and their properties*, Amer. Math. Soc., Providence, Rhode Island, 1983.

[2] R.A. AIRAPETJAN - G.M. HENKIN, *Integral representations of differential forms on Cauchy-Riemann manifolds and the theory of CR-functions*, Usp. Mat. Nauk **39** (1984), 39-106, [Engl. trans. Russ. Math. Surv. **39** (1984), 41-118], and: *Integral representations of differential forms on Cauchy-Riemann manifolds and the theory of CR-functions II*, Matem. Sbornik **127** (**169**) (1985), 1, [Engl. trans. Math. USSR Sbornik **55** (1986), 1, 91-111].

[3] A. ANDREOTTI - G. FREDRICKS - M. NACINOVICH, *On the absence of Poincaré lemma in tangential Cauchy-Riemann complexes*, Ann. Scuola Norm. Sup. Pisa, Cl. Sci. **8**, (1981), 365-404.

[4] A. ANDREOTTI - C.D. HILL, *E.E. Levi convexity and the Hans Lewy problem. Part I: Reduction to vanishing theorems*, Ann. Scuola Norm. Sup. Pisa, Cl. Sci. **26**, (1972), 325-363.

[5] A. ANDREOTTI - C.D. HILL, *E.E. Levi convexity and the Hans Lewy problem. Part II: Vanishing theorems*, Ann. Scuola Norm. Sup. Pisa, Cl. Sci. **26**, (1972), 747-806.

[6] M.Y. BARKATOU, *Estimation Höldérienne d'un noyau local de type Martinelli-Bochner sur les hypersurfaces 1-convexes-concaves. Applications*, Prépublication de l'Institut Fourier n° 250, Grenoble, 1993, to appear in Math. Z.

[7] B. FISCHER - J. LEITERER, *A local Martinelli-Bochner formula on hypersurfaces*, Math. Z. **214** (1993), 659-681.

[8] B. FISCHER - J. LEITERER, *A jump theorem with uniform estimates for $\bar{\partial}_b$-closed forms on real hypersurfaces*, Math. Nachr, **162** (1993), 109-116.

[9] G.M. HENKIN, *The Lewy equation and analysis on pseudoconvex manifolds (russ.)*, Usp. Mat. Nauk **32** (1977), 57-118, [Engl. trans. Russ. Math. Surv. **32** (1977), 59-130].

[10] G.M. HENKIN, *Solution des équations de Cauchy-Riemann tangentielles sur des variétés de Cauchy-Riemann q-convexes*, C.R. Acad. Sci. Paris Sér. I Math. **292** (1981), 27-30.

[11] G.M. HENKIN, *Analytic representation for CR-functions on submanifolds of codimension 2 in $\mathbb{C}^n$*, Lecture Notes in Math. Springer **798** (1980), 169-191.

[12] G.M. HENKIN, *The method of integral representations in complex analysis (russ.)*. In: Sovremennge problemy matematiki, Fundamentalnye napravlenija, Moscow Viniti **7** (1985), 23-124, [Engl. trans. in: Encyclopedia of Math. Sci., Several complex variables I, Springer-Verlag, **7** (1990), 19-116].

[13] G.M. HENKIN, *The Hartogs-Bochner effect on CR manifolds*, Soviet. Math. Dokl. **29** (1984), 78-82.

[14] G.M. HENKIN - J. LEITERER, *Theory of functions on complex manifolds*, Akademie-Verlag Berlin and Birkhäuser-Verlag Boston, 1984.

[15] G.M. HENKIN - J. LEITERER, *Andreotti-Grauert theory by integral formulas*, Akademie-Verlag Berlin and Birkhäuser-Verlag Boston (Progress in Math. 74), 1988.

[16] C.D. HILL - M. NACINOVICH, *Pseudoconcave CR-manifolds, preprint*, Dipartimento di Mathematica, Pisa, **1.76 (723)**, 1993.

[17] C.D. HILL - M. NACINOVICH, *Aneurysms of pseudoconcave CR-manifolds, preprint*, Dipartimento di Mathematica, Pisa, **1.83 (745)**, 1993.

[18] J.J. KOHN - H. ROSSI, *On the extension of holomorphic functions from the boundary of a complex manifold*, Ann. of Math. **81** (1965), 451-472.

[19] C. LAURENT-THIEBAUT, *Résolution du $\overline{\partial}_b$ à support compact et phénomène de Hartogs-Bochner dans les variétés CR*, Proc. Sympos. Pure Math. **52** (1991), 239-249.

[20] C. LAURENT-THIEBAUT, *Phénomème de Hartogs-Bochner relatif dans une hypersurface réelle 2-concave d'une variété analytique complexe*, Math. Z. **212** (1993), 511-525.

[21] C. LAURENT-THIEBAUT - J. LEITERER, *Uniform estimates for the Cauchy-Riemann equation on q-convex wedges*, Ann. Inst. Fourier (Grenoble) **43** (1993), 383-436.

[22] C. LAURENT-THIEBAUT - J. LEITERER, *Uniform estimates for the Cauchy-Riemann equation on q-concave wedges*, Astérisque **217** (1993), 151-182.

[23] C. LAURENT-THIEBAUT - J. LEITERER, *On the Cauchy-Riemann equation on q-concave wedges*, Prépublication de l'Institut Fourier n° 274, Grenoble, 1994, to appear in Math. Pachrichten.

[24] C. LAURENT-THIEBAUT - J. LEITERER, *Théorie d'Andreotti-Grauert pour les hypersurfaces réelles d'une variété analytique complexe*, C. R. Acad. Sci. Paris Sér. I Math. **316** (1993), 891-894.

[25] C. LAURENT-THIEBAUT - J. LEITERER, *On the Hartogs-Bochner extension phenomenon for differential forms*, Math. Ann. **284** (1989), 103-119.

[26] C. LAURENT-THIEBAUT - J. LEITERER, *The Andreotti-Vesentini separation theorem with C^k estimates and extension of CR-forms, several complex variables*, Proceeding of the Mittag-Leffler Institute **1987-88**, edited by J.E. Fornaess, Math. Notes **38**, Princeton University Press (1993), 416-439.

[27] M. NACINOVICH, *Poincaré lemma for tangential Cauchy-Riemann complexes*, Math. Ann. **268** (1984), 449-471.

[28] M. NACINOVICH, *On strict Levi q-convexity and q-concavity on domains with piecewise smooth boundaries*, Math. Ann. **281** (1988), 459-482.

[29] M.C. SHAW, *Homotopy formulas for $\overline{\partial}_b$ in CR manifolds with mixed Levi signatures*, to appear.

[30] F. TREVES, *Homotopy formulas in the tangengial Cauchy-Riemann complex*, Mem. Ann. Math. Soc., **434**, 1990.

"Pantograf" - Via alla Stazione di Voltri 2/A - Genova
Finito di stampare nel Dicembre 1995